REVISE BTEC NATIONAL
Engineering

REVISION GUIDE

Series Consultant: Harry Smith

Authors: Andrew Buckenham, Kevin Metcalf, David Midgley and Neil Wooliscroft

---

A note from the publisher

While the publishers have made every attempt to ensure that advice on the qualification and its assessment is accurate, the official specification and associated assessment guidance materials are the only authoritative source of information and should always be referred to for definitive guidance.

This qualification is reviewed on a regular basis and may be updated in the future. Any such updates that affect the content of this Revision Guide will be outlined at **www.pearsonfe.co.uk/BTECchanges**. The eBook version of this Revision Guide will also be updated to reflect the latest guidance as soon as possible.

For the full range of Pearson revision titles across KS2, KS3, GCSE, Functional Skills, AS/A Level and BTEC visit:

www.pearsonschools.co.uk/revise

Pearson

Introduction

Which units should you revise?

This Revision Guide has been designed to support you in preparing for the externally assessed units of your course. Remember that you won't necessarily be studying all the units included here – it will depend on the qualification you are taking.

BTEC National Qualification	Externally assessed units
For each of: Extended Certificate Foundation Diploma Diploma	1 Engineeering Principles 3 Engineering Product Design and Manufacture
Extended Diploma	1 Engineering Principles 3 Engineering Product Design and Manufacture 6 Microcontroller Systems for Engineers

Your Revision Guide

Each unit in this Revision Guide contains two types of pages, shown below.

Content pages help you revise the essential content you need to know for each unit.

Skills pages help you prepare for your exam or assessed task. Skills pages have a coloured edge and are shaded in the table of contents.

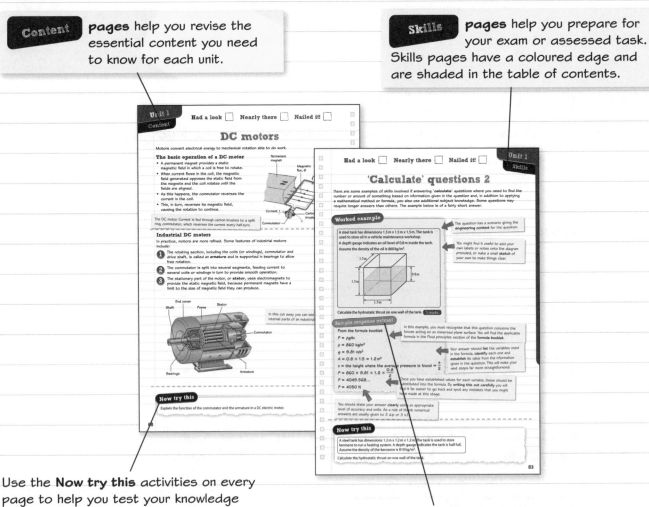

Use the **Now try this** activities on every page to help you test your knowledge and practise the relevant skills.

Look out for the **sample response extracts** to exam questions or set tasks on the skills pages. Post-its will explain their strengths and weaknesses.

Contents

Unit 6: Microcontroller Systems for Engineers

A small bit of small print
Pearson publishes Sample Assessment Material and the Specification on its website. This is the official content and this book should be used in conjunction with it.
The questions in *Now try this* have been written to help you test your knowledge and skills. Remember: the real assessment may not look like this.

Laws of indices

The laws of indices make it possible to simplify and solve equations that contain indices.

Using the laws of indices

Numbers 1 to 3 below are the main laws of indices. Numbers 4 to 8 are useful applications and special cases of the main laws.

1 Multiplication $a^m \times a^n = a^{(m+n)}$

e.g. $x^6 \times x^{-2} = x^4$

2 Division $a^m \div a^n = a^{(m-n)}$

e.g. $x^5 \div x^3 = x^2$

3 Powers $(a^m)^n = a^{mn}$

e.g. $(x^3)^5 = x^{15}$

4 Reciprocals $\dfrac{1}{a^m} = a^{-m}$

e.g. $\dfrac{1}{x^5} = x^{-5}$

5 Index $= 0$ $\quad a^0 = 1$

e.g. $7^0 = 1$

6 Index $= \dfrac{1}{2}$ or 0.5 $\quad a^{\frac{1}{2}} = a^{0.5} = \sqrt{a}$

e.g. $16^{-\frac{1}{2}} = 16^{0.5} = \sqrt{16} = 4$

7 Index $= \dfrac{1}{n}$ $\quad a^{\frac{1}{n}} = \sqrt[n]{a}$

e.g. $27^{-\frac{1}{3}} = \sqrt[3]{27} = 3$

8 Index $= 1$ $\quad a^1 = a$

e.g. $3.5^1 = 3.5$

Start at the beginning of the expression and look for terms that can be expressed as decimals to make simplification easier. e.g. $a^{\frac{3}{2}} = a^{-1.5}$.

Group together like terms.

Add the indices for like terms.

Consider alternative ways to express the simplified expression.

Make all the indices decimals.

Group together like terms.

Add the indices for like terms.

Consider alternative ways to express the simplfied expression.

Worked example

(a) Simplify the expression $a^{-2}b^{-1}a^{\frac{3}{2}}$.

$a^{-2}b^{-1}a^{\frac{3}{2}} = a^{-2}b^{-1}a^{1.5}$

$= a^{-2}a^{1.5}b^{-1}$

$= a^{-0.5}b^{-1}$

$= \dfrac{1}{b\sqrt{a}}$

(b) Simplify the expression $\dfrac{a^{-1}b^{-1}b^{\frac{1}{2}}}{a^{-2}b}$.

$= a^{-1}b^{-1}b^{0.5}a^2b^{-1}$

$= a^{-1}a^2b^{-1}b^{0.5}b^{-1}$

$= a^1b^{-1.5}$

$= \dfrac{a}{b^{1.5}} = \dfrac{a}{b\sqrt{b}}$

Information Booklet of Formulae and Constants

In your Unit 1 exam, you will be given an Information Booklet of Formulae and Constants and this includes the multiplication, division and powers laws of indices. Ideally, you should be confident in their use without reference to them.

The booklet is included in this Revision Guide on pages 91 to 95.

Now try this

1 Evaluate
 (a) 19^0 (b) $16^{-\frac{1}{2}}$ (c) $64^{\frac{2}{3}}$ (d) $(\frac{1}{4})^{-2}$

2 Express $\sqrt{(x^a \times x^b)}$ as a power of x

3 Evaluate $\sqrt[3]{9} \times \sqrt[6]{9}$

Logarithms

Logarithms (or **logs**) are a way of writing facts about **powers**.

Logs

These two statements mean the same thing:

$$\log_a b = x \leftrightarrow a^x = b$$

You say, 'log to the base a of b equals x.'

$\log_a b = x \longleftrightarrow a^x = b$

a is the base of the logarithm.

Remembering the order

The key to being confident in log questions is remembering the *basic definition*. Start at the base, and work in a circle:

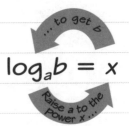

... to get b

$$\log_a b = x$$

Raise a to the power x ...

Laws of logarithms

Learn these four key laws for manipulating expressions involving logs. These laws work for all logarithms with *the same base*.

1 $\log_a x + \log_a y = \log_a(xy)$

    $\log_4 8 + \log_4 2 = \log_4 16 = 2$

2 $\log_a x - \log_a y = \log_a\left(\dfrac{x}{y}\right)$

    $\log_9 18 - \log_9 6 = \log_9 3 = \frac{1}{2}$

3 $\log_a\left(\dfrac{1}{x}\right) = -\log_a x$

    $\log_8\left(\dfrac{1}{x}\right) = -\log_8 2 = -\frac{1}{3}$

4 $\log_a(x^n) = n\log_a x$

    $\log_5(25^3) = 3\log_5 25 = 3 \times 2 = 6$

Worked example

Find:

(a) the positive value of x such that $\log_x 49 = 2$.

$x^2 = 49$

$x = 7$

(b) the value of y such that $\log_5 y = -2$.

$5^{-2} = y$

$y = \frac{1}{25}$

Changing the base

You can change the base of a logarithm using this formula:

$$\log_a x = \frac{\log_b x}{\log_b a} \qquad \log_9 27 = \frac{\log_3 27}{\log_3 9} = \frac{3}{2}$$

Common logs and natural logs

✓ Logarithms to base 10 are called common logs and the notation $\lg n$ is sometimes used instead of $\log_{10} n$; e.g. $\log_{10} 100 = 2$ (note $100 = 10^2$).

✓ Logarithms to base e are called natural logs with the notation $\log_e n$.

Now try this

Write down the corresponding power fact. Work in a circle, starting at the base.

1 Find:

   (a) the value of y such that $\log_3 y = -1$

   (b) the value of p such that $\log_p 8 = 3$

   (c) the value of $\log_4 8$.

2 Express as a single logarithm to base a.

   (a) $2\log_a 5$

   (b) $\log_a 2 + \log_a 9$

   (c) $3\log_a 4 - \log_a 8$

Use law 4 to write $3\log_a 4$ as $\log_a(4^3)$, then use law 1 to combine the two logarithms.

Exponential function

An exponential function (a^x) is one where the variable is the power, not the base. The Euler constant form of this expression (e^x) is found in many engineering disciplines such as aerodynamics, mechanics and electrical principles.

Exponential growth

If a population doubles or trebles every year, then it is said to be growing 'exponentially'.

We represent a population y that starts at 1000 and which doubles every year as $y = 1000 \times (2^x)$, where x represents the number of years.

🔗 Links You may need to use the laws of indices and logarithms when solving problems based on exponential growth or decay, see pages 1 and 2.

Worked example

Predict the number of people (y) after 4 years for a population of 500 that is trebling every year.

$y = 500 \times (3^4) = 500 \times 81 = 40\,500$

Using a calculator

Most calculators have the x^y button; for example, to find 3^4, press:

and the calculator should show 81.

Exponential function e^x

The Euler notation is a special form of the exponential function, written e^x. The exponential function e^x more than doubles, but does not quite treble over a period. The graph opposite shows $y = 2^x$, 3^x and e^x.

- A decaying population, such as radiated heat energy or radioactive particles, has the form $y = e^{-x}$.

- Recognise where the laws of indices can be used: $y = e^{-x}$ may also be written
$$y = \frac{1}{e^x}$$

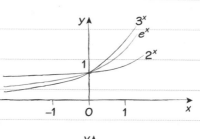

Graph of $y = 2^x$, $y = 3^x$ and $y = 3^x$

For $x = 0$, 2^x, 3^x and e^x are all equal to 1. For e^x the gradient at any point on the curve $y = e^x$ is equal to e^x.

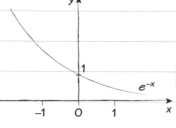

Graph of $y = e^{-x}$

Worked example

In a production process involving heat transfer, the temperature $\theta\,°C$ of a mould, at time t minutes, is given by $\theta = 250 + 150e^{-0.15t}$. Determine the temperature of the mould after 5 minutes.

$\theta = 250 + 150e^{-(0.15 \times 5)} = 250 + 150e^{-(0.75)}$

$\quad = 250 + 70.8549$

$\theta = 320°C$ (to 2 s.f.)

Using a calculator

Most calculators have the exponential button e^x button; it may be a secondary function of the 'ln' button. For example, to find $e^{2.5}$, press:

and the calculator should show 12.182....

Now try this

1 A manufacturer quadruples its production of a component from 2000 units per year every year for three years. Calculate the number of components produced at the end of the third year.

2 The voltage (V_c) across a capacitor in a RC circuit is given by $V_c = V_s(1 - e^{-\frac{t}{\tau}})$, where ($\tau$) is the time constant and V_s is the supply voltage. Determine the value of V_c at $t = 5\tau$ when the supply voltage is 4.5 V.

3

Equations of lines

The equation of a straight line can be written in the form $y = mx + c$, where m is the **gradient** of the line, and c is the point where it crosses the y-axis.

Point and gradient

 If you are given the gradient m of a straight line that passes through a point (x_1, y_1), then you can write its equation as:

$y - y_1 = m(x - x_1)$ to obtain an expression of y in terms of x.

Worked example

A straight line passes through the point $(-3, 2)$ and has a gradient -2. Find an equation for this line in the form $ax + by + c = 0$, where a, b and c are integers.

$$y - y_1 = m(x - x_1)$$
$$y - 2 = -2(x - (-3))$$
$$y - 2 = -2x - 6$$
$$y + 2x + 4 = 0$$

 If you are given two points on a line, (x_1, y_1) and (x_2, y_2), you can calculate the gradient using:
$$m = \frac{y_2 - y_1}{x_2 - x_1}$$

Worked example

The line L passes through the points $(1, 1)$ and $(2, 4)$. Find an equation for L in the form $y = mx + c$.

Gradient $(m) = \frac{4-1}{2-1} = 3$

$y = mx + c$

$y = 3x + c$

$1 = 3(1) + c$ (from point $(1, 1)$)

$c = -2$ so $y = 3x - 2$

Check using point $(2, 4)$:

$4 = 3(2) - 2$

Once you've evaluated the value of c, you can substitute this in the general equation of the straight line $y = mx + c$

Intercepts

You can find where the line $y = mx + c$ intercepts both the x- and the y-axes.

The y- intercept is given by the value of c, and the x- intercept can be evaluated by setting the value of y to 0.

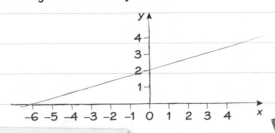

Graph of $y = \frac{1}{3}x + 2$

Worked example

Determine the intercepts for the x- and y-axes of the line $3y - x = 6$.

Rearranging the expression in the form $y = mx + c$ gives $y = \frac{1}{3}x + 2$, therefore the y-axis intercept is $+2$ (when $x = 0$).

Setting $y = 0$:

$\frac{1}{3}x + 2 = 0$

$x + 6 = 0$

$x = -6$ Therefore, the x- intercept is -6.

You can sketch a graph to check your answer. The gradient is positive because the m term $(\frac{1}{3})$ is positive and the line passes through the x-axis at -6 and the y-axis at $+2$.

Now try this

1. The line L passes through the point $(6, -5)$ and has gradient $-\frac{1}{3}$. Find an equation for L in the form $ax + by + c = 0$, where a, b and c are integers.

2. The line L passes through $(-4, 2)$ and $(8, 11)$. Find an equation for L in the form $y = mx + c$, where m and c are constants.

Simultaneous linear equations

Linear equations have the form $y = mx + c$ (i.e. no x^2 or y^2). Simultaneous equations can be solved using either the substitution or the elimination method. Whichever method you use, remember to *number* the equations to keep track of your working.

Substitution

Solve the simultaneous equations:

$y - 2x = 2$ Call this equation (1).

$-2y + 5 = x$ Call this equation (2).

$y = 2x + 2$ Call this equation (3).

Rearrange the linear equation (1) to make y the subject.

$-2(2x + 2) + 5 = x$

$-4x - 4 + 5 = x$

Substitute equation (3) into equation (2) and simplify to find x.

$5x = 1$

$x = \frac{1}{5} = 0.2$

You have found x. Substitute $x = 0.2$ into equation (1) to find y.

The solutions are $x = 0.2$, $y = 2.4$.

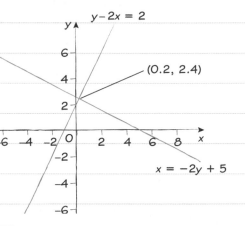

Thinking graphically

✓ The solutions to a pair of linear simultaneous equations correspond to the point where the graphs of the equations intersect.

✓ The point of intersection has an x value and a y value.

Graph showing simultaneous linear equations
$y - 2x = 2$ and $-2y + 5 = x$.

Worked example

You can substitute for x or y. It is easier to substitute for y because there will be no fractions.

Remember to number your equations.

Solve the simultaneous equations:

$y - 3x = 8$ (1)

$2y + 11 = -9x$ (2)

From (1): $y = 3x + 8$ (3)

Substitute (3) into (2) and simplify to find x:

$2(3x + 8) + 11 = -9x$

$6x + 16 + 11 = -9x$

$15x = -27$, $x = -\frac{27}{15} = -1.8$

Substitute $x = 1.8$ into equation (1) to find y:

$y - 3(-1.8) = 8$, $y = 8 - 5.4 = 2.6$

Remember that the value of x and y represent the coordinates of the point where the simultaneous equations intersect.

Elimination

Manipulate one of the equations to make either the x or the y terms exactly the same in both equations.

Worked example

Solve the simultaneous equations

$6x + 6 = 5y$ (1)

$3y + 2x = 7$ (2)

Multiply equation (2) by -3 and rearrange to make the x terms the same.

$-6x + 21 = 9y$ (3)

Add equation (1) to equation (3) to eliminate the term $6x$ and $-6x$.

$27 = 14y$, $y = 1.93$; now substitute into (2) to obtain x:

$(3 \times 1.93) + 2x = 7$, $x = 0.61$

You can check your solution by substituting $x = 0.61$ into equation (1):

$(6 \times 0.61) + 6 = 5y$

$y = 1.93$

Now try this

Solve the following simultaneous equations:

1 $2x + 13 = -3.5y$

    $-3x = -9y$

2 $14 = 3y + 5x$

    $10x = 4y + 7$

Expanding and factorisation

Expanding brackets and factorising expressions are the basis for manipulation of engineering expressions and formulae and are therefore essential skills.

Expanding brackets

To expand the product of two factors you have to multiply EVERY TERM in the first factor by EVERY TERM in the second factor:

There are 2 terms in the first factor and 3 terms in the second factor, so there will be $2 \times 3 = 6$ terms in the expanded expression BEFORE you collect like terms.

$2x \times x$
$3 \times x$

$$(2x + 3)(5x^2 - x + 4) = 10x^3 - 2x^2 + 8x + 15x^2 - 3x + 12$$
$$= 10x^3 + 13x^2 + 5x + 12$$

Simplify your expression by collecting like terms: $-2x^2 + 15x^2 = 13x^2$

Factorising

1 For simple expressions you can extract the Highest Common Factor (HCF).

2 Some expressions may provide a common factor within the bracketed terms.

$(3x - 2)$ is common to both terms so can be extracted as a factor.

$$2x(3x - 2) - 7(3x - 2) = (3x - 2)(2x - 7)$$

Terms OUTSIDE the brackets form the second factor $(2x - 7)$.

3 In some cases it may be necessary to group the terms to identify a common factor.

Start by grouping the first two and last two terms. Then extract the HCF as in part (a) above for each group, which leaves a common factor: $(2a - 3)$.

You can then extract the common factor as in 2 above.

Special cases

There are some special cases that you should watch out for:

✓ Completing the square (See page 7 for how this works.)
$x^2 + 2bx + c = (x + b)^2 - b^2 + c$

✓ The difference of two squares
$(a + b)(a - b) = a^2 - b^2$

Assuming no resistance, you can calculate the distance travelled by an object, using $s = ut + \frac{1}{2}at^2$.

Worked example

(a) Factorise the RHS of this equation.

$s = t(u + \frac{1}{2}at)$

Take out the common factor in each term. Don't forget to check your answer by multiplying back out. You should get the original expression.

(b) Factorise $3a(x + 4a) + 2(x + 4a)$.

$3a(x + 4a) + (x + 4a) = (x + 4a)(3a + 2)$

$(x + 4a)$ is common to both terms so can be taken as one of the factors. $(3a + 2)$ are the terms outside the brackets gathered together and which form the second factor.

(c) Factorise $6a^2 - 9a - 4ab + 6b$.

$(6a^2 - 9a) - (4ab - 6b)$

$3a(2a - 3) - 2b(2a - 3)$

$= (2a - 3)(3a - 2b)$

Now try this

Hint: to factorise $V_j^2 - V^2$, use the difference of two squares.

1 Expand the brackets to show that $(3x - 4)(3x + 4) = 3x^2 - 16$.

2 Factorise $3x(y - 4) - 2(y - 4)$.

3 The thrust to speed efficiency of a jet engine is given by the expression $\frac{3V(V_j - V^2)}{V_j^2 - V^2}$.

Factorise and, hence, show that the complete expression is equal to $\frac{3V}{V_j + V}$.

Quadratic equations 1

Quadratic equations occur throughout engineering in different forms. You must be able to identify them and know how to solve them using three methods. Factorisation and completing the square are shown below. Use of the formula is shown on page 8. You will find practical uses for these methods on page 20.

Factorising a quadratic

You can follow these steps to solve some quadratic equations:

1 **Rearrange** the equation into the form $ax^2 + bx + c = 0$.

2 **Factorise** the left-hand side.

3 Set each factor equal to zero and solve to find two values of t: $(t - 3)(t + 5) = 0$.

> The valid solution is the value of t that is greater than 0. This is given by the factor $(t - 3)$, therefore the answer is $t = 3\,\text{s}$.
>
> You could indicate this in your answer by writing: where $t \geq 0$.

Worked example

The displacement of a car in metres (s) is given by $s = 2t + t^2$, where t is in seconds. Find how long it takes to travel $15\,\text{m}$.

Find two numbers with a sum of +2 and a product of −15.

$t^2 + 2t - 15 = 0$

$(t - 3)(t + 5) = 0$ ——— The required numbers are '−3' and '+5'.

$t - 3 = 0$ or $t + 5 = 0$

$t = 3\,\text{s}$ or $t = -5\,\text{s}$

Discount the negative root in which $t = -5\,\text{s}$.

Completing the square

Useful for quadratics that do not factorise.

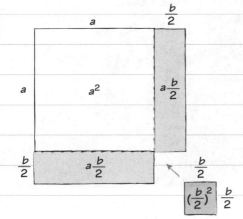

Complete the square with $(\frac{b}{2}) \times (\frac{b}{2})$; i.e. $(\frac{b}{2})^2$

The complete square is made of three terms:

1 $a^2 + ab + (\frac{b}{2})^2$

2 These are simplified as $(a + \frac{b}{2})^2$.

3 Remember to add $(\frac{b}{2})^2$ to both sides of the equation.

Worked example

The area of a rectangular building (length x) is given by $4 = x(5 - x)$, where the width in metres is $(5 - x)$. Find the roots and, hence, the length and width.

$x^2 - 5x = -4$

$x^2 - 5x + (\frac{5}{2})^2 = -4 + (\frac{5}{2})^2$

Complete the square by adding $(\frac{5}{2})^2$ to both sides.

$(x - 2.5)^2 = 2.5^2 - 4$

$x - 2.5 = \pm\sqrt{2.25}$

therefore $x = +1.5 + 2.5 = 4$ (length)

or $x = -1.5 + 2.5 = 1$ (width)

$x^2 - 5x + (\frac{5}{2})^2$ may be written as $(a + \frac{b}{2})^2$ or $(x - 2.5)^2$.

The coefficient of the x term is the numerator of the 'complete the square' term.

Now try this

The distance x, in metres, along a beam where the bending moment $= 0$ is given by $5x^2 + 14x - 3$. Factorise this expression and, hence, find the position of x.

You need to put this expression $= 0$, find the roots and then discard negative values. Remember to include the units (metres).

Quadratic equations 2

In some cases, you will be unable to solve quadratic equations by factorisation or completing the square. You will need to solve them using the formula instead. You will find practical uses for this method on page 81.

Solution by formula

You can solve any quadratic by use of the formula, but it *must* be in the form $ax^2 + bx + c = 0$, where:

$$x = \frac{-b \pm \sqrt{b^2 - 4ac}}{2a}$$

c is the 'c' term

a is the coefficient of x^2

b is the coefficient of *x*

Quadratics come in many forms, for example:
$$120l = 10l^2 + 100$$
$$(y - 2)^2 = 18$$
You need to recognise the different forms and use the appropriate method to find the solution.

Using the formula

The formula will be on the formulae sheet, but be confident in using the discriminant to check the nature of the roots.

✓ If $b^2 - 4ac > 0$ then there are two solutions.

✓ If $b^2 - 4ac = 0$ then the quadratic has one solution.

✓ If $b^2 - 4ac < 0$ then the quadratic doesn't have real roots.

You will not be asked to solve quadratics where $b^2 - 4ac < 0$.

Worked example

A duct manufacturer produces a rectangular ducting sheet of 28 m² in which the area is related to the width by the expression $28 = 1.6w^2 + 4.2w$. What is the width of the sheet?

$$28 = 1.6w^2 + 4.2w \text{ or } 1.6w^2 + 4.2w - 28 = 0$$

$a = 1.6, b = 4.2, c = -28$

$$w = \frac{-4.2 \pm \sqrt{4.2^2 - 4 \times (1.6 \times (-28))}}{2 \times 1.6}$$

$$w = \frac{-4.2 \pm 14.029}{3.2} = 3.07 \text{ or } -5.69$$

Reject negative answer, width = 3.07 m (to 2 d.p.)

Robotic arm on a Mars lander. Calculating the distance to turn a robotic arm in mid-motion is one use of the quadratic formula to solve $s = ut + \frac{1}{2}at^2$.

Now try this

The height, *h*, of a ball thrown vertically is given by
$$h = -4.3t^2 + 54t + 13$$
where *t* is time, measured in seconds. The time to reach the ground will be given when $h = 0$. Calculate the time taken for the ball to reach the ground, using the quadratic equation.

The equation will provide two solutions. In this example, one of them will be negative, which should be rejected.

Don't forget to specify the units.

Radians, arcs and sectors

Radians are an alternative unit of angular measurement to degrees. They must be used in calculations involving **circular motion** and **rotational dynamics**.

Radians

1 radian is the angle formed when the radius of a circle is 'wrapped' around the circumference. There are 2π radians in a full circle. Use these rules to convert between radians and degrees:

🔗 **Links** There is more about angular motion on page 27.

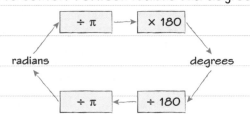

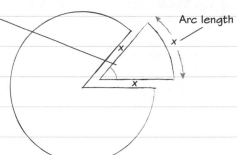

This sector has an arc length equal to its radius. The angle at the centre of the circle is 1 radian. You can also write '1 rad', or just '1'.

Arc length

Arc length and sector area

You can calculate the area of a sector and the length of an arc easily if the angle is measured in radians.

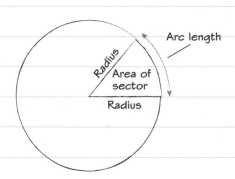

Arc length

Area of sector

Radius

1 Length of an arc

$s = r\theta$

2 Area of a sector

$A = \frac{1}{2} r^2 \theta$

Both these formulae are given in the formulae booklet. They work **only** if the angle θ is measured in **radians**.

Worked example

A microchip manufacturer uses silicon wafers in the shape of a sector of a circle, with the dimensions shown.

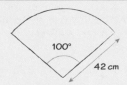

100°

42 cm

Calculate the area of one wafer.

$100° = \frac{100}{180} \times \pi = \frac{5}{9} \pi$ rad

Area $= \frac{1}{2} r^2 \theta = \frac{1}{2} \times 42^2 \times \frac{5}{9} \pi = 1539.3804$

$\approx 1540\,cm^2$ (to 3 s.f.)

Common angles

Save time converting between radians and degrees by learning these five common angles:

Degrees	45	60	90	180	360
Radians	$\frac{\pi}{4}$	$\frac{\pi}{3}$	$\frac{\pi}{2}$	π	2π

Convert the angle to radians then use the formula for the area of the sector. Remember to round your final answer to 3 significant figures and give the correct units.

Now try this

1 Find the arc length and area of the sector of a circle, with radius 4 cm, which contains an angle of 30°.

2 A plasma cutter is used to cut sectors of a circle for ventilation trunking. The arc length of each sector is 450 mm and the radius is 1 m. Find the angle, in radians, of a sector and, hence, the number of complete sectors that can be obtained from a circle of sheet metal with radius 1 m.

Trigonometric ratios and graphs

You need to be able to recall the trigonometric identities of SIN, COS and TAN and their respective graphs.

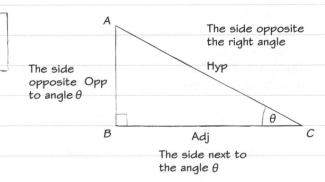

Basic trigonometric ratios

✓ $\sin \theta = \frac{Opp}{Hyp}$ ✓ $\cos \theta = \frac{Adj}{Hyp}$ ✓ $\tan \theta = \frac{Opp}{Adj}$

You can use the mnemonic 'SOH CAH TOA' to remember these ratios. (Sine: Opposite over Hypotenuse, Cosine: Adjacent over Hypotenuse, Tangent: Opposite over Adjacent.)

The side opposite the right angle

Hyp

The side opposite Opp to angle θ

The side next to the angle θ

Adj

$y = \sin x$ and $y = \cos x$

$y = \tan x$

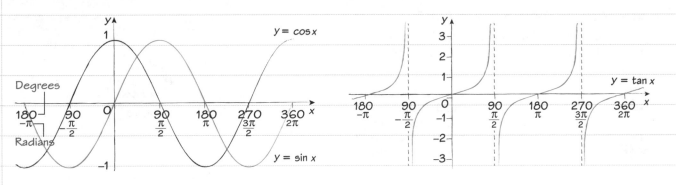

Trig values for θ

The value of θ (pronounced theta) may be represented in degrees, or as radians, in terms of π.

θ (°)	θ (radians)	$\sin \theta$	$\cos \theta$	$\tan \theta$
0	0	0	1	0
45	$\frac{\pi}{4}$	0.707	0.707	1
90	$\frac{\pi}{2}$	1	0	$-\infty$
270	$\frac{3\pi}{2}$	−1	0	$-\infty$
360	2π	0	1	0

Make sure you are confident using both 'rad' and 'deg' modes on your calculator.

Now try this

1   Produce a table that states the values of $\sin \theta$, $\cos \theta$ and $\tan \theta$ at the following intervals: 0°, 30°, 45°, 60°, 90°, 135°, 180°, 270°, 360°. Include a column in the table for the radian equivalent of each of these angles.

2   Evaluate the length of BC in triangle ABC, in which angle B is a right-angle, angle A is $\frac{\pi}{4}$ rad and AB is 10 cm.

First make a rough sketch of the triangle.

Cosine rule

The cosine rule applies to *any* triangle. You usually use the cosine rule when you know two sides and the included angle (SAS) or when you are given three sides and you want to work out an angle (SSS). Although the cosine rule will be provided in the formulae sheet in your exam, you need to be confident in its use.

 1 $a^2 = b^2 + c^2 - 2bc \cos A$

This version is useful for finding a missing side.

 2 $\cos A = \dfrac{b^2 + c^2 - a^2}{2bc}$

Use this version to find a missing angle.

Using a calculator

If you're using the sine rule or cosine rule you need to make sure that your CALCULATOR is in the correct mode (DEGREES or RADIANS).

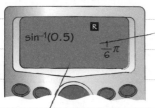

$\sin^{-1}(0.5)$
$\frac{1}{6}\pi$

This calculator is in radians mode.

On some calculators you need to press SHIFT and SETUP to change between degrees and radians mode.

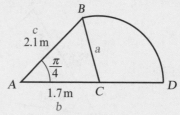

Worked example

The diagram shows a triangle, *ABC*, and a sector, *BCD*, of a circle with centre *C*. Find the length of *BC*.

$a^2 = b^2 + c^2 - 2bc \cos A$

$BC^2 = 1.7^2 + 2.1^2 - 2 \times 1.7 \times 2.1 \times \cos\frac{\pi}{4}$

$= 2.2512\ldots$

$BC = 1.50\,\text{m}$ (to 2 d.p.)

SAS → cosine rule. If the angle on the diagram is given in terms of π then it is in **radians**. Make sure your calculator is set to radians before working out $\cos\frac{\pi}{4}$.

Worked example

In the triangle *ABC*, *AB* = 15 cm, *BC* = 9 cm and *CA* = 10 cm. Find the size of angle *C*, giving your answer in radians to 2 decimal places.

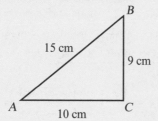

$\cos C = \dfrac{BC^2 + AC^2 - AB^2}{2 \times BC \times AC}$

$= \dfrac{9^2 + 10^2 - 15^2}{2 \times 9 \times 10}$

$C = \cos^{-1}(-0.2444\ldots) = 1.82\,\text{rad}$ (to 2 d.p.)

SSS → cosine rule. Be careful with the order. You add the squares of the sides **adjacent** to the angle, and subtract the square of the side **opposite**.

Now try this

The diagram shows two triangles, *PQR* and *PRS*. ∠ *RSP* = 0.8 radians. Find:

(a) the length of *PR*

(b) the size of ∠ *PQR*, giving your answer in radians to 3 significant figures.

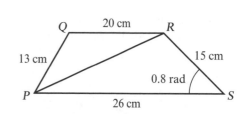

Sine rule

The sine rule applies to *any triangle*. The sine rule is useful when you know *two angles*, or when you know a side and the *opposite* angle. Although the sine rule will be provided in the formulae sheet in your exam, you need to be confident in its use.

① $\dfrac{a}{\sin A} = \dfrac{b}{\sin B} = \dfrac{c}{\sin C}$

This version is useful for finding a missing side.

② $\dfrac{\sin A}{a} = \dfrac{\sin B}{b} = \dfrac{\sin C}{c}$

Use this version to find a missing angle.

Worked example

In the triangle *PQR*,
$PQ = 4$ cm,
$\angle PQR = 1.5$ radians
and
$\angle QPR = 0.8$ radians.
Find the length of the side *PR*.

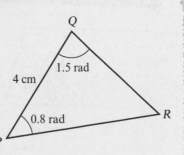

$\angle QRP = \pi - 1.5 - 0.8 = 0.8415\ldots$

$\dfrac{PR}{\sin(1.5\ \text{rad})} = \dfrac{4}{\sin(0.8415\ldots\ \text{rad})}$

$PR = \dfrac{4 \times \sin(1.5\ \text{rad})}{\sin(0.8415\ldots\ \text{rad})} = 5.35$ cm (to 2 d.p..)

Using radians

Remember these key facts:

✓ Angles in a triangle add up to π radians.

✓ A right angle is $\dfrac{\pi}{2}$ radians.

🔗 **Links**   There is more about measuring angles in radians on page 9.

The sine rule compares **opposite** sides and angles, so use the fact that the angles in a triangle add up to π radians (or 180°) to work out $\angle QRP$ first.

Two values

If you know $\sin x$, you might be asked to find *two possible values* for x. Use this rule for angles measured in *radians*:

$\sin x = \sin(\pi - x)$

You could draw a sketch to help you see what is going on.

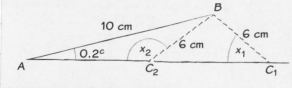

Worked example

In the triangle *ABC*, $AB = 10$ cm, $BC = 6$ cm, $\angle BAC = 0.2$ radians and $\angle ACB = x$ radians.

(a) Find $\sin x$, giving your answer to 2 decimal places.

$\dfrac{\sin x}{10} = \dfrac{\sin(0.2\ \text{rad})}{6}$

$\sin x = \dfrac{10 \times \sin(0.2\ \text{rad})}{6} = 0.33$ (to 2 d.p.)

(b) Given that there are two possible values of x, find these values of x, correct to 2 decimal places.

$x_1 = \sin^{-1}(0.331) = 0.34$ radians (2 d.p.)

$x_2 = \pi - x_1 = \pi - 0.34 = 2.80$ radians (2 d.p.)

Now try this

In the triangle *ABC*, $BC = 13$ cm, $\angle ABC = \dfrac{\pi}{3}$ radians, and $\angle ACB = \dfrac{\pi}{5}$ radians. Find the length of *AC*.

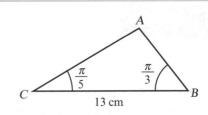

Start by finding the size of $\angle CAB$.

Vector addition

Vectors such as force, velocity and ac phasors have both magnitude and direction. You need to take both into account when finding the overall resultant of two or more vectors.

Visualising vectors

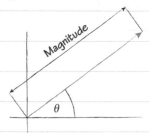

Here the magnitude of the vector is proportional to the length of the arrow. The direction is given by the angle (θ) and the sense by the arrowhead.

Vector diagrams

A vector diagram helps you visualise a system where two (or more) vector quantities are acting and find their sum or resultant vector.

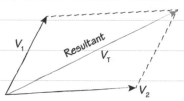

Vector diagram using the parallelogram law to show the resultant of two vectors.

Graphical vector addition

To find the resultant of a system of vectors:

- Draw them accurately and to scale on graph paper.
- Complete the parallelogram and draw in the diagonal that represents the resultant.
- Measure the resultant using a ruler and protractor.

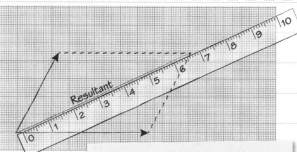

Taking accurate measurements to determine the magnitude and direction of a resultant vector.

Analytical vector addition

A more accurate solution can be found using trigonometry to find the resultant of two vectors.

🔗 **Links**   You will find analytical methods of vector addition in specific applications are covered on page 16 Resolving forces and page 72 Addition of sinusoidal waveforms.

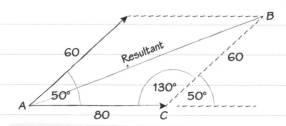

Here you can apply the cosine rule to calculate the resultant magnitude and direction.

🔗 **Links**   You will find the cosine rule on page 11 .

Now try this

The analysis of a system of forces produces the vector diagram shown. The diagram is not to scale.

Find, using a graphical approach, the magnitude and direction of the resultant vector.

Compare your graphical solution with that found using the analytical approach described above.

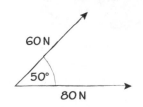

Surface area and volume

You need to know how to calculate the surface areas and volumes of cylinders, spheres and cones.

Cylinder

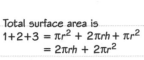
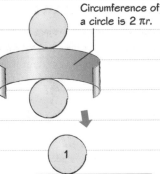
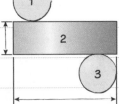

Area of a circle is $\pi r^2 h$.

Circumference of a circle is $2\pi r$.

Total surface area is
$1+2+3 = \pi r^2 + 2\pi rh + \pi r^2$
$= 2\pi rh + 2\pi r^2$

Volume $= \pi r^2 h$, Total surface area $= 2\pi rh + 2\pi r^2$

Worked example

A weapon release unit on an aircraft is protected by a cover in the shape of a cylinder of radius 40 mm and height 60 mm, open at one end. Determine the total surface area of the component.

The total surface area of a complete cylinder $= 2\pi rh + 2\pi r^2$, but one end is open.

Total surface area of component

$= 2\pi rh + \pi r^2$

$= (2\pi \times 0.04 \times 0.06) + (\pi \times 0.04^2)$

$= 2.01 \times 10^{-2}$ m² (to 2 d.p.)

Make sure that units are consistent. Here all measurements are converted to metres.

Sphere

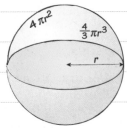

$4\pi r^2$ $\frac{4}{3}\pi r^3$ r

Surface area $= 4\pi r^2$
Volume $= \frac{4}{3}\pi r^3$

Worked example

Calculate the minimum number of portable fuel pods, in the shape of 3 metre diameter polymer spheres, that are required to supply a town in a disaster-struck area, with 25 000 litres of fuel (1 m³ = 1 × 10³ litres).

Volume of one fuel pod

$= \frac{4}{3}\pi r^3 = \frac{4}{3}\pi \left(\frac{3}{2}\right)^3 = 9.42$ m³ $= 9420$ litres

Number of fuel pods $= 25\,000 \div 9420 = 2.65$

A minimum of 3 fuel pods are required.

Cone

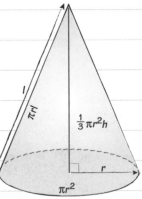

l πrl $\frac{1}{3}\pi r^2 h$ r πr^2

Surface area $= \pi rl + \pi r^2$
Volume $= \frac{1}{3}\pi r^2 h$

Worked example

A food manufacturer wishes to market a new confectionary product, in the shape of a cone. Calculate the height (h) of the product if each cone contains 41.85 cm³ of filling and has a diameter of 4.0 cm.

Volume $= \frac{1}{3}\pi r^2 h$

$41.85 \times 10^{-6} = \frac{1}{3}\pi \times 0.02^2 \times h$

$h = 0.10$ m or 10.0 cm (to 2 d.p.)

If the question doesn't specify which units to use for the answer then you can decide. Here all measurements are converted into metres.

Now try this

1 A cylindrical concrete piling has radius 0.7 m and height 12 m. Calculate the volume of concrete required.

2 If you double the radius of a sphere, how does the volume of the new sphere compare with the volume of the original sphere?

First make a sketch with the known information.

Systems of forces

Mechanical systems can contain components that exert forces that push or pull other components. In a static system all the forces are balanced so, instead of a force causing a change in the motion of an object, the system remains at rest or in motion with constant velocity.

Space diagrams

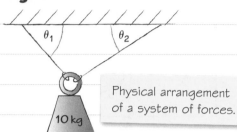

Physical arrangement of a system of forces.

Free body diagrams

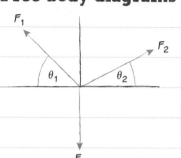

You use a free body diagram to show only the forces acting in the system you are investigating.

Coplanar concurrent forces

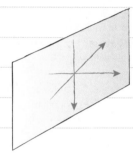

Coplanar concurrent forces act in a single plane and pass through a common point.

The resultant of coplanar concurrent forces

You can simplify coplanar concurrent forces to a single **resultant** force.

An **equilibrant** force can be used to balance the **resultant**. It has the same magnitude and direction but the opposite sense.

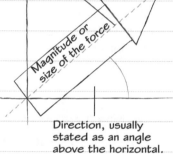

Sense, usually with left to right as the positive direction.

Direction, usually stated as an angle above the horizontal.

Fully defined resultant of a coplanar concurrent system of forces stating magnitude, direction and sense.

Coplanar non-concurrent forces

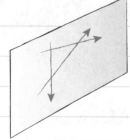

Coplanar non-concurrent forces act in a single plane but do not all pass through a common point.

The resultant of coplanar non-concurrent forces

You can also simplify coplanar **non**-concurrent forces to a single resultant. However, these commonly involve turning effects and so an extra piece of information is required to fully define them.

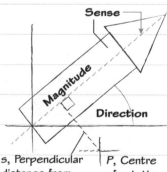

Sense

Direction

s, Perpendicular distance from the line of action of the force

P, Centre of rotation

Fully defined resultant force of a non-concurrent system with magnitude, direction, sense and distance from a centre of rotation to the line of action.

Now try this

The space diagram shows a stationary weight on a slope or inclined plane. Friction is preventing the weight from sliding down the slope.

Draw a free body diagram for this system.

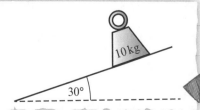

Note that, in your exam, you may find the terms 'free body diagram' and 'space diagram' are used interchangeably.

Resolving forces

To easily add forces together they must be acting in the same direction. First resolve them into **horizontal** and **vertical components**. Then add or subtract to find the resultant.

Formulae for resolving forces

Imagine a right-angled triangle with the force as its hypotenuse. For a force of magnitude F acting at an angle θ to the horizontal:

$F_v = F \sin \theta$

$F_h = F \cos \theta$

$\dfrac{F_v}{F_h} = \tan \theta$

$F^2 = F_v^2 + F_h^2$

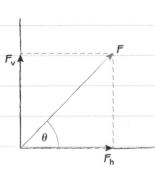

Worked example

The diagram shows coplanar forces acting on a demolition ball as it swings through the air.

Calculate the **magnitude** and **direction** of the resultant force.

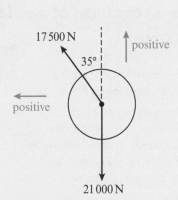

Resolving horizontally:

$17\,500 \sin 35° = 10\,037.58... \text{ N}$

Resolving vertically:

$17\,500 \cos 35° - 21\,000 = -6664.83... \text{ N}$

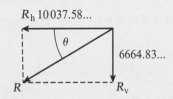

$R = \sqrt{10\,037.58...^2 + 6664.83...^2} = 12\,000\,\text{N (3 s.f.)}$

$\tan\theta = \dfrac{6664.83...}{10\,037.58...}$

$\theta = 33.6° \text{ (3 s.f.)}$

Choose a **positive direction** that makes your working easiest. Here the positive direction is **up**. The resultant vertical force is **negative**, so it acts downwards.

1. Resolve the 17 500 N force into vertical and horizontal components.
2. Find the **resultant** vertical and horizontal forces acting on the system.
3. **Sketch** a triangle with the overall resultant force as the hypotenuse.
4. Use Pythagoras' theorem to find the magnitude of the resultant.
5. Use trigonometry to find the direction of the resultant.

An engineer could resolve the tensions in the strings to work out the resultant force acting on this hot-air balloon basket.

Now try this

The diagram represents the tension forces acting at a single point in a structural framework.

Calculate the magnitude and direction of the resultant for this system of coplanar forces.

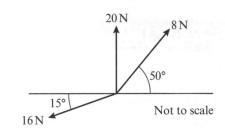

Not to scale

Moments and equilibrium

A moment is a measure of the **turning effect** of a force on a body. It is found from the **magnitude** of the force and the **perpendicular distance** from the line of action of the force to the centre of rotation.

Turning moment (*M*)

$$M = Fs$$

Turning moment (newton metres (N m))

Force, in newtons (N)

Perpendicular distance from the line of action of the force to the centre of rotation (s)

Turning moments can be used to describe the turning force applied to a nut or bolt head by a spanner.

The turning effect or moment depends on the applied force and the length of the handle.

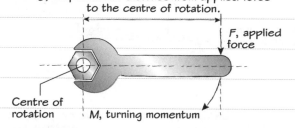

s, Perpendicular distance from applied force to the centre of rotation.

F, applied force

Centre of rotation

M, turning momentum

Equilibrium

For a system to be in **static equilibrium** all forces and turning moments must balance each other.

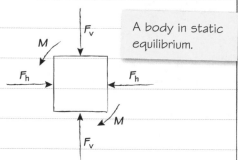

A body in static equilibrium.

The following three conditions must all be satisfied:

1 The sum of the horizontal components is 0.

$$\Sigma F_h = A_h + B_h \ldots + n_h = 0$$

2 The sum of the vertical components is 0.

$$\Sigma F_v = A_v + B_v \ldots + n_v = 0$$

3 The sum of the moments is 0.

$$\Sigma M = M_A + M_B \ldots + M_n = 0$$

Worked example

The diagram shows a simple system of forces acting on a beam that pivots at point *A* and is being held in static equilibrium by an equilibrant force *E*.

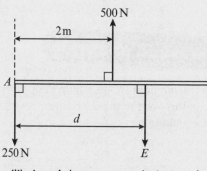

Calculate the magnitude, direction and sense of force *E* and distance *d*.

Consider the forces in equilibrium (where upwards is +ve):

$$0 = 500 - 250 - E$$

$$E = 500 - 250$$

$E = 250$ N acting vertically with −ve (downward) sense

Consider moments in equilibrium about A (where clockwise is +ve):

$$0 = (250 \times 0) + (d \times 250) - (2 \times 500)$$

$$d = \frac{1000}{250}$$

$d = 4$ m from point A

Now try this

Determine whether the system of forces acting on this square plate is in static equilibrium by finding the sum of the vertical and horizontal components of the forces present and taking moments about point *A*.

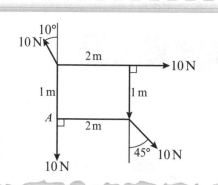

 Links To revise resolving forces and resolving a force into horizontal and vertical components, see page 16.

17

Simply supported beams

Simply supported beams are commonly used in engineering structures such as bridges, vehicle chassis and as supports over door and window openings in buildings. As an engineer, you must be able to analyse the forces required to support each end of a beam to make sure the supports are strong enough.

A simply supported beam

C_v equivalent point load replacing UDL acting at its centre

D_v point load

Uniformly distributed load (UDL)

A_h horizontal support reaction

A_v Vertical support reaction

B_v Vertical support reaction

Several different types of forces can act on a simply supported beam. You may be asked to calculate the support reactions for a given system.

Types of beam supports

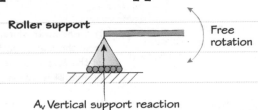

Roller support

Free rotation

A_v Vertical support reaction

Pin support

A_h Horizontal support reaction

Free rotation

A_h Vertical support reaction

You might also be asked to consider different types of beam supports that are able to provide support reaction forces in the vertical or horizontal directions.

Worked example

The diagram shows a simply supported beam as part of the plans to build a bridge. The beam is in static equilibrium.

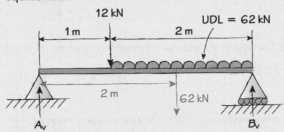

12 kN

1 m   2 m

UDL = 62 kN

2 m

62 kN

A_v B_v

Calculate the support reactions A_v and B_v.

Take moments about A to find B_v:

$(1 \times 12) + (2 \times 62) = (3 \times B_v)$

$B_v = 45.33$ kN (to 2 d.p.)

Take moments about B to find A_v:

$(3 \times A_v) = (1 \times 62) + (2 \times 12)$

$A_v = 28.67$ kN (to 2 d.p.)

You can consider the uniformly distributed load (UDL) as a point load acting at the centre at the distribution. In this case, that would mean a point force acting vertically downwards 2 m from support A. Add this to the diagram.

1 Replace any uniformly distributed loads by equivalent point loads.

2 Take moments about support A. Anti-clockwise moments will equal clockwise moments, as the beam is in equilibrium.

3 Rearrange to find B_v.

4 Take moments about support B.

5 Rearrange to find A_v.

6 Check your answer.

Now try this

Check the solution in the Worked example by determining whether the beam still satisfies all the conditions of static equilibrium.

Refer back to page 17 Moments and equilibrium to find the conditions that must be met for static equilibrium.

 Links You could also work through the additional beam problem given on page 78.

Direct loading

Direct loading includes tensile forces, which pull and stretch a component, and compressive forces, which push and squeeze a component. Direct loading gives rise to direct stress and direct strain.

Direct stress (σ)

Direct stress is a measure of the direct load distribution within a material.

$$\text{Direct stress } (\sigma) = \frac{\text{Normal force } (F)}{\text{Area } (A_\sigma)}$$

Stress has units N/m^2 or Pa ($N/m^2 = 1\,Pa$).

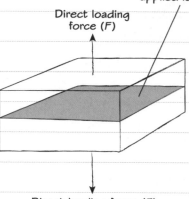

Force (F) is distributed over an area (A_σ) inside the material perpendicular to the applied load.

Direct loading force (F)

Direct loading force (F)

Direct strain (ε)

Direct strain is a measure of the deformation caused by an applied direct stress.

$$\text{Direct strain } (\varepsilon) = \frac{\text{Change in length } (\Delta L)}{\text{Original length } (L)}$$

Strain is a dimensionless quantity and has no units.

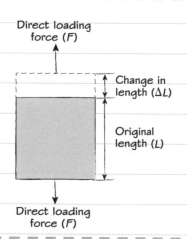

Direct loading force (F)

Change in length (ΔL)

Original length (L)

Direct loading force (F)

Also known as Young's modulus, the Modulus of elasticity (E) expresses the linear relationship between direct stress and direct strain.

$$\text{Modulus of elasticity } (E) = \frac{\text{Direct stress } (\sigma)}{\text{Direct strain } (\varepsilon)}$$

Modulus of elasticity has units N/m^2 or Pa.

Now try this

The sketch shows part of a structural beam that is loaded in tension.
Calculate the direct stress in the beam.

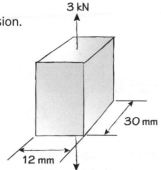

3 kN

30 mm

12 mm

Shear loading

Shear loading is caused by forces that cut across a component and tend to shear or cut it apart. Shear loading gives rise to **shear stress** and **shear strain**.

Shear stress (τ)

Shear stress is a measure of the shear load distribution within a material.

$$\text{Shear stress } (\tau) = \frac{\text{Shear force } (F)}{\text{Shear area } (A_\tau)}$$

Stress has units N/m² or Pa (N/m² = 1 Pa).

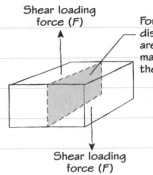

Shear loading force (F)

Force (F) is distributed over an area (A_τ) inside the material parallel to the applied load.

Shear loading force (F)

Shear strain (γ)

Shear strain is a measure of the deformation caused by an applied shear stress.

$$\text{Shear strain } (\gamma) = \frac{\text{Change in length } (\Delta L)}{\text{Original length } (L)}$$

Strain is a dimensionless quantity and has no units.

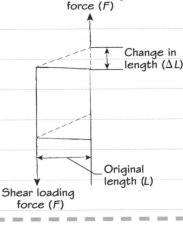

Shear loading force (F)

Change in length (ΔL)

Original length (L)

Shear loading force (F)

Rigidity

The modulus of rigidity (G) expresses the linear relationship between shear stress and shear strain.

$$\text{Modulus of rigidity } (G) = \frac{\text{Shear stress } (\tau)}{\text{Shear strain } (\gamma)}$$

Modulus of rigidity has units N/m² or Pa.

Now try this

Rivets are often used to join two parallel metal plates. A cross-section of such an arrangement is shown in the diagram.

Calculate the shear stress in the rivet.

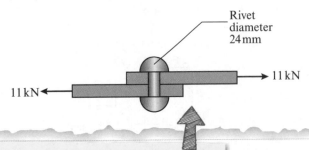

Rivet diameter 24 mm

11 kN

11 kN

Don't forget to change the values given in the question to standard units before performing any calculations.

Find the cross-sectional area of the rivet.

Velocity, displacement and acceleration

Engineers use the SUVAT equations to analyse the motion of objects travelling in straight lines with constant acceleration. Although they will be on the formulae sheet in your exam, you need to be confident in using and manipulating them.

Constant acceleration formulae

Here are the first two formulae you need to use for motion in a straight line with constant acceleration:

① $v = u + at$

② $s = \dfrac{(u+v)}{2} t$

Displacement (s)
Initial velocity (u)
Final velocity (v)
Acceleration (a)
Time (t)

Look at the diagram on the right to see what each letter represents.

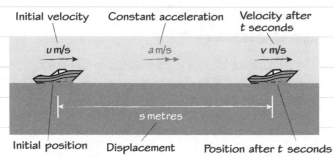

Initial velocity   Constant acceleration   Velocity after t seconds

u m/s   a m/s   v m/s

s metres

Initial position   Displacement   Position after t seconds

Using SUVAT

The constant acceleration formulae are sometimes called the SUVAT formulae. In the exam you should write down all five letters.

☑ Write in any values you KNOW.

☑ Put a QUESTION MARK next to the value you want to find.

☑ CROSS OUT any values you don't need for that question.

This will help you choose which formula to use.

Read the question carefully. The distance AB is 1.2 km, but the aircraft does not take off at B.

If you have to solve a constant acceleration question involving three points like this one, it's a good idea to draw a quick sketch to help you see what is going on. In part (b), s is the distance AC, in metres. You need to subtract it from 1200 m to find the distance CB.

Worked example

When taking off, an aircraft moves on a straight runway, *AB*, of length 1.2 km. The aircraft moves from *A* with initial speed 2 m/s. It moves with constant acceleration and 20 s later it leaves the runway at *C* with speed 74 m/s. Find:

(a) the acceleration of the aircraft

$s = ?, u = 2, v = 74, a = ?, t = 20$

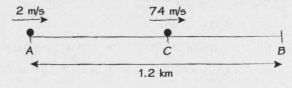

2 m/s   74 m/s

A   C   B

1.2 km

$v = u + at$

$74 = 2 + a \times 20$

$20a = 72$

$a = 3.6\,\text{m/s}^2$

(b) the distance *CB*.

$s = ?, u = 2, v = 74, a = 3.6, t = 20$

$s = \dfrac{(u+v)}{2} t$

$AC = \dfrac{(2+74)}{2} \times 20$

$= 760\,\text{m}$

$CB = 1200\,\text{m} - 760\,\text{m} = 440\,\text{m}$

Now try this

A car moves along a straight stretch of road *AB*. The car moves with initial speed 2 m/s at point *A*. It accelerates constantly for 12 seconds, reaching a speed of 23 m/s at point *B*. Find:

(a) the acceleration of the car

(b) the distance *AB*.

You can answer both parts of this question using the formulae given on this page. But you can use any of the other SUVAT formulae if you are confident with them.

Applying the SUVAT equations

If you know three of s, u, v, a and t, you can calculate the missing values.

① $v^2 = u^2 + 2as$

② $s = ut + \dfrac{1}{2}at^2$

③ $s = vt - \dfrac{1}{2}at^2$

For an understanding of what each letter represents, look at the blue box.

Worked example

A train moves along a straight track with constant acceleration. Three telegraph poles are set at equal intervals beside the track at points A, B and C, where $AB = 50$ m and $BC = 50$ m. The front of the train passes A with speed 22.5 m/s, and 2 s later it passes B.

Find:

(a) the acceleration of the train

$s = 50$, $u = 22.5$, $v = ?$, $a = ?$, $t = 2$

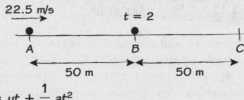

$s = ut + \dfrac{1}{2}at^2$

$50 = 22.5 \times 2 + \frac{1}{2} \times a \times 2^2$

$50 = 45 + 2a$

$a = 2.5 \, \text{m/s}^2$

(b) the speed of the front of the train when it passes C.

$s = 100$, $u = 22.5$, $v = ?$, $a = 2.5$, $t = ?$

$v^2 = u^2 + 2as$

$v^2 = 22.5^2 + 2 \times 2.5 \times 100$

$\quad = 1006.25$

$v = 31.721... = 31.72 \, \text{m/s}$ (to 2 d.p.)

Units

You need to make sure that your measurements are in the correct units.

✓ t (time) is measured in seconds.

✓ s (displacement) is measured in metres.

✓ u, v (velocity) is measured in m/s.

You will sometimes see ms^{-1} written as m/s. They both mean the same thing: metres per second.

✓ a (acceleration) is measured in m/s^2.

You will sometimes see m/s^2 written as ms^{-2}. They both mean the same thing: metres per second squared, or metres per second per second.

(b) This formula involves v^2, so there are two possible values of v. You have been asked to find the speed of the front of the train. Speed is the magnitude of the velocity, so you need to give a positive answer.

Now try this

A boat travels in a straight line with constant deceleration, between two buoys, A and B, 300 m apart. The boat passes buoy A with initial speed 16 m/s, and passes buoy B 30 seconds later. Find:

(a) the deceleration of the boat

(b) the speed of the boat as it passes B.

Deceleration is represented in the SUVAT equations by 'a' but remember to make it a **negative** value by putting a '−' in front of it.

Force, friction and torque

You need to understand what is meant by force, friction and torque and be able to calculate their values in different contexts.

Force

A force is a push or a pull acting on an object. It is measured in newtons (N).

Friction

Friction can be considered as two distinct cases:

- Static frictional force: opposes a static object from starting to move when a force is applied
- Kinetic frictional force: opposes a moving object's motion.

Calculating frictional forces

To calculate frictional forces, you need a value for the resistance to movement provided by different surfaces:

☑ Coefficient of static friction: $F_s = \mu_s N$, where F_s is the limiting frictional force (in newtons) and N is the normal reaction between the surfaces (mass × g, in newtons)

☑ Coefficient of kinetic friction: $F_k = \mu_k N$, where F_k is the kinetic frictional force (in newtons) and N is the normal reaction between the surfaces (mass × g, in newtons).

Kinetic friction is often referred to as dynamic or sliding friction.

Worked example

Determine the torque acting on a bolt when a force of 150 N is applied perpendicularly to a spanner with length 30 cm.

$\tau = ?$, $F = 150\,\text{N}$, $r = 0.30\,\text{m}$

$\tau = Fr$

$= 150 \times 0.30 = 45\,\text{N m}$

Remember that torque is the force applied multiplied by the distance from the centre of the object.

Worked example

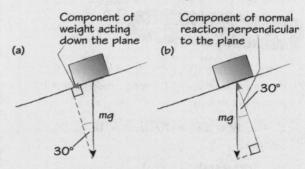

A 30° inclined plane is used to reduce the force required to raise a load of 12 kg, where the coefficient of kinetic friction is 0.4.

(a) Calculate the component of the weight acting down the plane.

(b) Calculate the component of normal reaction acting perpendicular to the plane.

(c) Calculate the kinetic frictional resistance.

(d) Calculate the total force acting down the plane.

(a) $mg \sin 30 = 12 \times 9.81 \times 0.5 = 58.8\,\text{N}$

(b) $mg \cos 30 = 12 \times 9.81 \times 0.87 = 101.84\,\text{N}$

(c) $F_k = \mu_k N = 0.4 \times 101.84 = 40.74\,\text{N}$

(d) $F = 58.8 + 40.74 = 99.54\,\text{N}$

Torque

The turning moment of a couple is called torque, represented by the Greek letter τ (pronounced tau).

It can be considered as a force (F) applied tangentially to a shaft or wheel of radius (r), where F is in newtons and r in metres.

Torque is the product of the force and the radius, and is measured in newton metres (N m): $\tau = Fr$.

Now try this

1 A 20° inclined plane is used to reduce the force required to raise a load of 50 kg, where the coefficient of kinetic friction is 0.35. Calculate the total force acting down the plane.

2 A 50 mm diameter bar is being turned on a lathe. The force on the cutting tool, which is tangential to the surface of the bar, is 0.7 kN. Calculate the applied torque.

Work and power

Work is force × distance and **power** is energy transferred in time taken.

Mechanical work

Work is done when force is exerted on an object and

- it moves a certain distance
- it moves in the direction of the applied force,

$$W = Fs$$

Work in joules (J), where 1 J is 1 N m.   Force, in newtons (N)   Distance, in metres (m)

- Take care with units: if the force is in kN then the work will be in kJ, not J.
- $F = ma$, or in this special case weight = mg.
- The direction of movement is in the direction of the applied force; vertically upwards.

Worked example

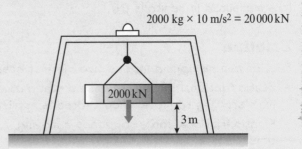

$2000 \text{ kg} \times 10 \text{ m/s}^2 = 20\,000 \text{ kN}$

2000 kN

3 m

A gantry hoist is used to lift a pallet with a mass of 2000 kg through a distance of 3 m. Taking $g = 10 \text{ m/s}^2$, calculate the work done.

$W = ?$ J, $F = 2000 \times 10$ N, $s = 3$ m

$W = Fs = 2000 \times 10 \times 3$ J

$\quad = 60\,000$ J (60 kJ)

Worked example

In the example above, the pallet was raised through 3 metres in 12 seconds. Calculate the power of the hoist.

$P = ?$, $W = 60 \times 10^3$ J, $t = 12$ seconds

$$\frac{W}{t} = \frac{60 \times 10^3}{12} \quad 5 \times 10^3 \text{ W} = 5 \text{ kW}$$

A selector ram pushes a component horizontally at an average velocity of 5.2 m/s with a force of 2450 N. Calculate the power output of the ram.

$W = ?$, $F = 2450$ N, $s = 5.2$ m, $t = 1$ s

$W = Fs = 2450 \times 5.2$

$\quad\quad = 12740$ J = 12.74 kJ

$$P = \frac{W}{t} = \frac{12.74 \text{ kJ}}{1 \text{ second}}$$

$\quad = 12.74$ kW

Power

Energy transferred may be in different forms; for example, heat transfer or lifting a weight.

$$\text{Power} = \frac{\text{Energy transferred}}{\text{Time taken}} = \frac{\text{Work done}}{\text{Time taken}} = \frac{W}{t}$$

The unit of power is the watt (W), 1 watt = 1 joule of energy transferred in 1 second.

Also, $P = F \times v$ v is the average velocity (i.e. $\frac{u+v}{t}$), in m/s.

F is the force, in N.

Average and instantaneous power

✓ If the energy transfer is averaged; for example, average speed of an object pushed by a force or temperature change of an object averaged over a period of time, then the power will be an 'averaged' value.

✓ The instantaneous power is the average power as the time factor approaches zero.

Now try this

1   A spring is extended by 5.0 cm using 100 J of work. Calculate the average force applied.

2   A car travels 105 metres along a straight road in 10 seconds at a constant velocity. Frictional forces are constant at 500 N. Calculate the power output of the engine, assuming 100% transmission system efficiency.

First find the amount of work done, then calculate the power using the time taken.

Energy

Energy can be thought of as being involved in doing work, making objects move and heating something up. In doing work the energy is not used up, it transfers from one energy store, such as gravitational potential energy, to others, such as kinetic energy, sound and heat.

Gravitational potential energy (GPE)

GPE depends *only* on the **mass** of the object, the **acceleration due to gravity** and the **height** of the object. It doesn't matter whether the object is moving or resting on another object.

GPE, in joules

$$E_p = mgh$$

Height, in metres

Mass, in kilograms   Acceleration due to gravity ($9.81\,\text{m/s}^2$)

Remember that $m \times g$ is the same as the weight of an object, which is a force, measured in newtons (N).

In calculations, convert to SI units and make sure you can manipulate very small as well as very large values.

Worked example

In a spray dry production facility, synthetic detergent granules are manufactured by ejecting liquid droplets at high speed from a rotating atomiser disc in the presence of hot air. The individual droplets have a mass of 1.30 milligrams and a GPE at point of ejection from the rotating disc of 51 μJ. Calculate the height of the atomiser disc above the base of the unit.

$E_p = 5.10 \times 10^{-5}\,\text{J}$, $m = 1.30 \times 10^{-6}\,\text{kg}$, $g = 9.81\,\text{m/s}^2$, $h = ?$

$$E_p = mgh$$

$$5.10 \times 10^{-5} = 1.30 \times 10^{-6} \times 9.81 \times h$$

$$h = 4.00\,\text{m (to 2 d.p.)}$$

Kinetic energy (KE)

Energy associated with motion such as falling, rotating or moving in a straight line.

$$E_k = \tfrac{1}{2}mv^2$$

v is the velocity, in m/s.

m is the mass, in kg

GPE decreases due to the reducing height.
KE increases due to acceleration under gravity.

🔗 **Links** Read more about the conservation of energy on page 26.

You may be given an energy transfer question in which it is assumed that all the GPE is converted to KE (see question 2 below). That is not the case here because the gondolas are almost perfectly balanced and are driven around using a small motor.

Worked example

Each 600 tonne gondola on the Falkirk Wheel contains 500 000 litres of water and takes 5.5 minutes to complete the 53 metre journey from the upper canal to the lower one. Assuming constant velocity throughout the travel, calculate the kinetic energy of one of the gondolas. Assume the mass of water in one gondola to be 500 000 kg.

$E_k = ?\,\text{J}$, $v = ?$, $s = 53\,\text{m}$, $t = (5.5 \times 60)\,\text{s}$, $m = (600 + 500)$ tonnes $= 1.1 \times 10^6\,\text{kg}$

$$E_k = \tfrac{1}{2}mv^2$$

$$v = \frac{d}{t} = \frac{53}{330} = 0.16\,\text{m/s (2 d.p.)}$$

$$E_k = \tfrac{1}{2}(1.1 \times 10^6 \times 0.16^2)$$

$$= 14\,080\,\text{J or 14.08 kJ}$$

Make sure that the units are correct; e.g. convert tonnes to kg.

Now try this

1. Using the data above, calculate the GPE of one of the gondolas on the Falkirk Wheel at its highest position (35 metres) above ground level.

2. A 200 kg pile driver, used to drive in reinforced concrete foundation piles, falls 7.5 metres under gravity. Ignoring the effects of air resistance, calculate its velocity at the point of impact with the concrete pile.

Newton's laws of motion, momentum and energy

Real-world engineering deals with objects in motion. You need to understand how to analyse these dynamic systems and the forces involved.

Newton's laws of motion

1 **Inertia:** objects continue in their state of rest, or of uniform velocity, as long as no net force acts. For rotating objects the moment of inertia (I) is related to the mass (m) and the radius of gyration (r): $I = mr^2$.

2 **Force:** the acceleration of an object is proportional to the size of an applied force and takes place in the direction of that force.

3 To every action there is an equal and opposite reaction.

> If at rest, a force is needed to start movement. If moving, a force is needed to stop it.

> Unbalanced forces lead to acceleration of an object in the direction of the resultant force. At constant mass: $F = ma$.

> Provides a sense of symmetry when considering the action of forces.

Momentum

Remember that momentum (p) is a vector quantity and is the product of mass and velocity, or $p = mv$ with units kg m/s.

Conservation of momentum

- ☑ Derives from Newton's third law
- ☑ Applies to 'isolated systems'
- ☑ Usually a collision or impact involved
- ☑ The sum of the momentum before the collision equals the sum of momentum after it.

> Note that change in momentum is called **impulse** which is also equal to force × time with units of newton seconds (N s).

Worked example

A 400 g piston in a diesel engine has an instantaneous momentum at mid-stroke of 2 kg m/s. Calculate its velocity.

$p = 2$ kg m/s, $m = 0.4$ kg, $v = ?$ m/s

$p = mv$, $\dfrac{p}{m} = \dfrac{2}{0.4} = 5$ m/s

A hammer of mass 0.75 kg with a velocity of 15 m/s hits a nail of mass 20 g. Assuming no rebound, what is the velocity of the hammer and nail immediately after the impact?

$m_h = 0.75$ kg, $v_h = 15$ m/s, $m_n = 20$ g, $v_{h+n} = ?$

$m_h \times v_h = m_{h+n} \times v_{h+n}$

$0.75 \times 15 = (0.75 + 0.02) \times v_{h+n}$

$v_{h+n} = \dfrac{11.25}{0.77} = 14.61$ m/s (to 2 d.p.)

Conservation of energy

- Energy is not lost when work is done.
- Energy can be considered to change from one energy 'store' to another, or from one form of energy to a different form. 'Energy can be transferred usefully, stored or dissipated. It cannot be created or destroyed.'

> Use a Sankey diagram to help you account for all energy changes. Energy in *must* equal energy out.

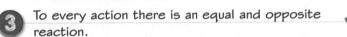

Total energy in / Useful energy out / Wasted energy out

Now try this

1 A mass of 130 000 kg moves along a straight, level track with a velocity of 15 m/s and collides with a second mass of 50 000 kg travelling in the same direction with a velocity of 10 m/s. After the collision the masses remain locked together. Calculate the velocity of the combined mass.

2 A car of mass 1250 kg stands on an incline of 7°. If the hand brake is released, calculate the velocity of the car after travelling 50 m down the incline. Assume the resistances to motion total 50 N.

Angular parameters

Engineering in the 'real world' often involves rotating objects, such as shafts, or objects moving along a curved path, for example, a F1 car cornering. Basic understanding of this motion and the power/energy calculations involved requires use of the formulae covered below.

Angular and linear velocity

Angular velocity (ω) is the same for all points on a rotating object; but **linear** (tangential) velocity (v) will depend on the distance from the centre (r): $v = r\omega$, where v is in m/s, r in metres and ω in rad/s.

 Links You can refresh your memory on radian measurement on page 9.

Centripetal acceleration

A point at distance r from the centre of a body rotating at constant angular velocity (ω) experiences an acceleration even though its speed is not changing. Its direction changes due to the acceleration a towards the centre of rotation: $a = \omega^2 r$ or $a = \dfrac{v^2}{r}$.

We are exploring only **centripetal acceleration** here, not the forces that would result from this acceleration being applied to a mass.

Worked example

A conveyer belt on a production line travels at 5.0 m/s and is driven by a pulley wheel of diameter 300 mm. Calculate the angular velocity.

$v = 5$ m/s, $r = 0.3$ m, $\omega = ?$ rad/s

$v = r\omega$

$5 = 0.3\omega$

$\omega = 16.67$ rad/s

Worked example

The pilot of a Naval F35 carries out a 1.5 km radius turn at a constant speed of 173 m/s.

(a) Determine the centripetal acceleration.

(b) Estimate the additional 'g' acceleration the pilot would experience as a result of the turn.

(a) $a = ?$ m/s^2, $v = 173$ m/s, $r = 1500$ m

$a = \dfrac{v^2}{r} = \dfrac{173^2}{1500} = 19.95$ m/s^2

(b) Acceleration due to gravity = 9.81 m/s^2, therefore 19.95 m/s^2 represents an additional acceleration of approximately 2g.

Power (P)

Power, the rate of work done (W) — $\boxed{P = \tau\omega}$ — Angular velocity (rad/s)

Torque (N m)

Kinetic energy (E_k)

Energy of a rotating body (J) — $\boxed{E_k = \tfrac{1}{2}I\omega^2}$ — Angular velocity (rad/s)

Moment of inertia (kg m^2)

Remember that $I = mr^2$, so this can be substituted in the formula because the mass and the radius of the rotating mass are given, whereas the moment of inertia is not.

Worked example

A constant torque of 70 N m keeps a pump rotating at 20 rev/s. Neglecting losses, find the power input to the pump.

$P = \tau\omega$
$P = 70 \times (20 \times 2\pi) = 8796$ W = 8.80 kW (2 d.p.)

Calculate the angular kinetic energy of a navigational gyroscope rotor, with mass 100 g and radius 40 mm, rotating at 100 rev/s.

$E_k = \tfrac{1}{2}I\omega^2 = \tfrac{1}{2}(mr^2) \times (100 \times 2\pi)^2$

$= \tfrac{1}{2}(0.1 \times 0.04^2) \times (394784.18)$

$= 31.58$ J

Now try this

1 A jet engine applies a constant torque of 850 N m to its turbofan at 100 rev/s. Calculate the power input to the turbofan.

2 A flywheel has a moment of inertia of 4 kg m^2 about its axis of rotation. Calculate the angular kinetic energy stored in the wheel when it is rotating at 8 rev/s.

Mechanical power transmission

Machines use design features such as levers, gears and screw threads, to make it easier to do work. Engineering projects would be impossible without using machines.

Mechanical advantage (MA)

Make sure you don't confuse load and effort:

1 **Load** is used for the Output force (F_l), measured in newtons (N).

2 **Effort** is used for the Input force (F_e), also measured in newtons.

$$MA = \frac{Load}{Effort}$$

- For a pulley system

  MA = the number of pulleys (providing the mass of the pulleys are ignored)

- For a lever

$$MA = \frac{Load}{Effort} = \frac{Effort\ to\ fulcrum\ distance}{Load\ to\ fulcrum\ distance}$$

Worked example

The force required to cut (shear) a branch with these pruning shears is 0.9 kN. Find the effort needed.

$$Effort = \frac{Load}{MA} = \frac{0.9 \times 10^3 \times 35 \times 10^{-3}}{145 \times 10^{-3}} = 217\,N$$

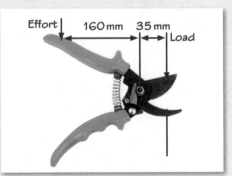

Effort | 160 mm | 35 mm | Load

Velocity ratio

Is the ratio of the distance moved by the effort and load.

$$VR = \frac{Distance\ moved\ by\ effort}{Distance\ moved\ by\ load}$$

or

$$VR = \frac{Velocity\ with\ which\ effort\ moves}{Velocity\ with\ which\ load\ moves}$$

For a pulley system, VR = the number of ropes supporting the load.

Worked example

A 180 kg load is lifted through 240 mm when the effort moves through 2.4 m. What is the velocity ratio?

$$VR = \frac{Distance\ moved\ by\ effort}{Distance\ moved\ by\ load} = \frac{2.4}{240 \times 10^{-3}}$$

$$VR = 10$$

Efficiency

Efficiency (η) (pronounced 'eta')

$$\eta = \frac{WD_l}{WD_e} \times 100\%$$

Useful work output — Work input

Efficiency also can be stated in terms of MA and VR:

$$\eta = \frac{MA}{VR}$$

Worked example

A machine has an efficiency of 65%. If an effort of 160 N raises a load of 1 tonne, find the velocity ratio.

$$\eta = \frac{MA}{VR}$$

$$MA = \frac{Load}{Effort} = \frac{1 \times 10^3 \times 9.81}{160} = 61.31$$

$$\eta = \frac{MA}{VR}\ therefore:$$

$$65 = \frac{61.31}{VR}$$

$$VR = \frac{65}{61.31} = 1.06$$

Now try this

1 A machine requires 7 kW of input power and the output power is 5.9 kW. Calculate the efficiency of the machine.

2 A machine lifts a load of 0.5 tonnes through a distance of 0.23 m when the effort of 300 N moves through a distance of 75 mm. Determine the VR, the MA and the efficiency.

Submerged surfaces

In applications such as dams and storage tanks, forces act on the submerged surfaces that retain the fluids. You need to understand these forces to ensure that the structures are built with sufficient strength to contain the fluids safely.

The pressure exerted on a submerged rectangular surface increases linearly as the submerged height, h, increases. (h is always measured from the surface downwards.)

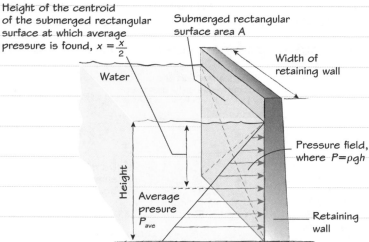

Height of the centroid of the submerged rectangular surface at which average pressure is found, $x = \frac{X}{2}$

Submerged rectangular surface area A

Width of retaining wall

Water

Height

Average presure P_{ave}

Pressure field, where $P = \rho g h$

Retaining wall

Hydrostatic pressure

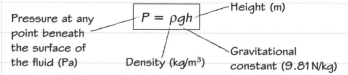

Pressure at any point beneath the surface of the fluid (Pa) — $P = \rho g h$ — Height (m)

Density (kg/m³)   Gravitational constant (9.81 N/kg)

Average hydrostatic pressure

The average pressure on a submerged rectangular plane surface acts at the height of its centroid:

$$P_{ave} = \rho g \frac{h}{2} = \rho g x$$

Hydrostatic thrust

Hydrostatic thrust F_T is the total force acting on the submerged plane surface and can be calculated as a function of average pressure and the area over which it is applied:

$$F_T = \rho g A x$$

Centre of pressure

The hydrostatic thrust can be thought of as a single point force acting at the centre of pressure. The height of the centre of pressure below the surface, hp, is given by the centroid of the triangular pressure field described as h increases.

$$hp = \frac{2}{3} h$$

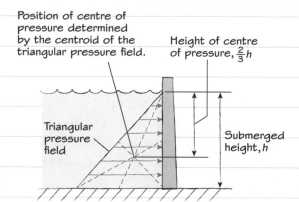

Position of centre of pressure determined by the centroid of the triangular pressure field.

Height of centre of pressure, $\frac{2}{3}h$

Triangular pressure field

Submerged height, h

A tank is used to store seawater. Calculate:

1 the hydrostatic thrust acting on the tank wall, illustrated in the diagram

2 the height of the centre of pressure.

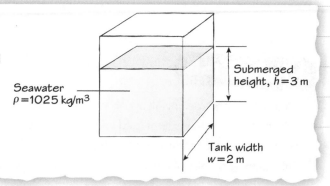

Submerged height, $h = 3$ m

Seawater $\rho = 1025$ kg/m³

Tank width $w = 2$ m

Immersed bodies

Archimedes' principle states that, 'A body totally or partially submerged in a fluid displaces a volume of fluid that weighs the same as the apparent loss in weight of the body.' A floating body experiences an upwards force equal in magnitude to its weight. A body that sinks experiences an apparent weight loss equal to the weight of the fluid it displaces.

Suspended body submerged in a fluid

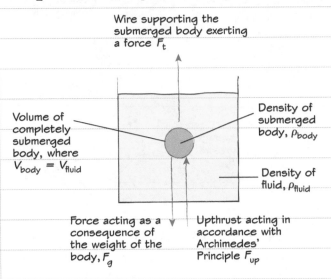

Wire supporting the submerged body exerting a force F_t

Volume of completely submerged body, where $V_{body} = V_{fluid}$

Density of submerged body, ρ_{body}

Density of fluid, ρ_{fluid}

Force acting as a consequence of the weight of the body, F_g

Upthrust acting in accordance with Archimedes' Principle F_{up}

Assuming that the object is in static equilibrium then the forces acting on the body are:

$$F_t = F_g - F_{up}$$
$$F_g = \rho_{body} V_{body} g$$
$$F_{up} = \rho_{fluid} V_{fluid} g$$

Floating bodies

A partially submerged floating body does not need to be suspended to maintain equilibrium.

When a body is floating and $F_t = 0$ then:

$$F_g = F_{up}$$
$$\rho_{fluid} V_{fluid} = \rho_{body} V_{body}$$

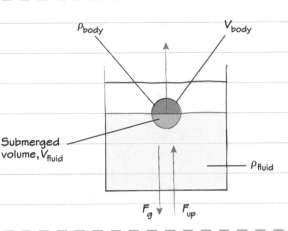

ρ_{body} V_{body}

Submerged volume, V_{fluid}

ρ_{fluid}

F_g F_{up}

Relative density

Relative density (d) is defined as the density of a substance compared to the density of pure water:

$$d_{substance} = \frac{\rho_{substance}}{\rho_{water}}$$

where $\rho_{substance}$ and ρ_{water} are absolute densities.

Determining density

You can use a flotation method to determine ρ_{body} as long as you know ρ_{fluid}, V_{fluid} and V_{body}.

For this block of wood floating in seawater:

$$\rho_{body} = \frac{\rho_{fluid} V_{fluid}}{V_{body}} = \frac{1025 \times 0.02}{0.04} = 512.5 \, \text{kg/m}^3$$

$\rho_{fluid} = 1025 \, \text{kg m}^{-3}$

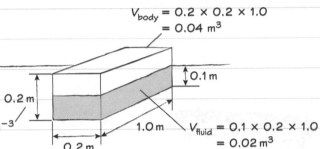

$V_{body} = 0.2 \times 0.2 \times 1.0$
$= 0.04 \, \text{m}^3$

0.1 m

0.2 m

1.0 m

0.2 m

$V_{fluid} = 0.1 \times 0.2 \times 1.0$
$= 0.02 \, \text{m}^3$

Now try this

A partially submerged plank of oak, with height 10 cm, width 25 cm and length 2.4 m, is floating in fresh water.

7.4 cm of the height is submerged under the water.

Fresh water has a density of 1000 kg/m³.

Calculate the density of the oak.

Fluid flow in tapering pipes

Tapering pipes can be used to alter the flow velocity of fluids travelling through them. You only need to revise **incompressible** fluids where density remains constant throughout the system.

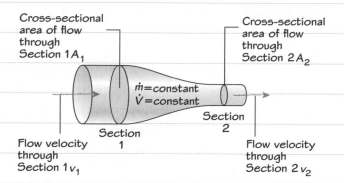

Cross-sectional area of flow through Section 1 A_1

Cross-sectional area of flow through Section 2 A_2

$\dot{m}$=constant
$\dot{V}$=constant

Section 1

Section 2

Flow velocity through Section 1 v_1

Flow velocity through Section 2 v_2

Volumetric flow rate ($\dot{V}$)

The volume (V) of fluid to pass a given point in time (t). Units m³/s:

$$\dot{V} = \frac{V}{t}$$

This can also be expressed in terms of flow velocity (v) and the cross-sectional area of the pipe (A):

$$\dot{V} = Av$$

Mass flow rate ($\dot{m}$)

The mass (m) of fluid to pass a given point in time (t). Units kg/s:

$$\dot{m} = \frac{m}{t}$$

This can also be expressed in terms of the density of the fluid (ρ) and the volumetric flow rate (V):

$$\dot{m} = \rho V$$

or

$$\dot{m} = \rho Av$$

Equations describing the continuity of flow

Since you are only dealing with incompressible fluids that maintain constant density throughout, then the volumetric and mass flow rates will remain constant throughout.

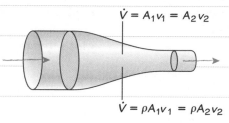

$$\dot{V} = A_1 v_1 = A_2 v_2$$

$$\dot{V} = \rho A_1 v_1 = \rho A_2 v_2$$

Always draw a labelled diagram when solving these types of problems.

Now try this

A hose has a gradually tapering nozzle that reduces the flow diameter from 25 mm to 10 mm. The flow velocity in the 25 mm diameter section is 5 m/s. Calculate the flow velocity after the nozzle.

Heat transfer parameters and thermal conductivity

Before you review heat transfer processes, check that you are familiar with the different parameters that will be used.

Temperature (T)

Temperature is a measure of the kinetic energy of atomic or molecular vibrations within a body.

The Celsius scale

Temperature is most often measured in degrees Celsius (°C).

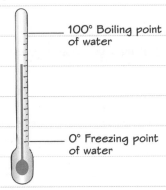

100° Boiling point of water

0° Freezing point of water

Thermodynamic temperature (T_k)

✓ In thermodynamic calculations, you must use the Kelvin (K) temperature scale.

✓ The zero point on the Kelvin scale is absolute zero (0 K).

✓ 0 K is the theoretical point at which all atomic or molecular vibrations cease. It cannot get any colder than absolute zero.

```
°C  ├────────────┼────────────┼────────────
   -273          0           100

K   ├────────────┼────────────┼────────────
    0           273          373
```

Pressure (P)

Pressure describes the action of a force (F) when distributed over an area (A). Units N/m² or Pascal (Pa).

Mass (m)

Mass is a measure of the amount of matter a body contains. It is distinctly different from weight as mass is independent of g. Units kg.

Time (t)

As engineers, you are most often interested in the rate at which change occurs. Many calculations are dependent upon time. Units s.

Linear dimensions

Many thermal processes cause expansion or contraction, which can be measured by changes in the linear dimensions of bodies; e.g. length, width, height, diameter. Units m.

Thermal conductivity (k)

Thermal conductivity is a material property that describes the ability of a material to conduct heat. Units W/m K.

Heat transfer rate is given by the equation:

$$Q = \frac{kA(T_a - T_b)}{x}$$

Often the surface finish of a component designed for rapid heat transfer is textured in order to maximise its surface area.

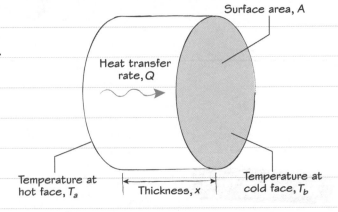

Surface area, A

Heat transfer rate, Q

Temperature at hot face, T_a

Thickness, x

Temperature at cold face, T_b

Now try this

Calculate the heat transfer rate through a 2 m² window, with an internal temperature on one side of 23 °C and an external temperature at the other of −4 °C, given the glass has thickness 3 mm and thermal conductivity 0.96 W/m K.

Heat transfer processes

Heat is transferred from one place to another by three distinct mechanisms: conduction, convection and radiation.

Conduction

Conduction involves the transmission of heat energy by:

- Atomic vibrations from one atom to another
- The movement of high-energy free electrons.

Conduction happens only when atoms are in contact, so gases tend to be poor thermal conductors.

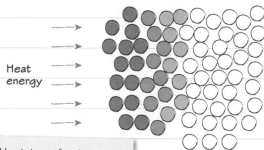

Heat energy

Heat transfer by conduction in a solid.

Convection

Convection can occur only in liquids or gases where the molecules are free to move.

Heating in one area of a fluid causes localised expansion and reduction in density.

Low density fluid rises, carrying heat energy with it.

Dense, low temperature fluid flows in to replace it and the cycle repeats.

A fan heater uses **forced convection** to distribute heat energy to its surroundings by blowing cool air over an electric heating element.

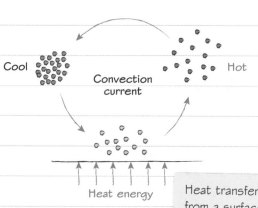

Cool   Convection current   Hot

Heat energy

Heat transfer away from a surface by a convection current.

Radiation

- No physical contact is required in heat transfer by radiation.
- Radiation does not need a medium through which to travel and can move through a vacuum.

Radiation involves the transmission of heat energy as electromagnetic waves released by high-energy electrons as they move between atomic orbits.

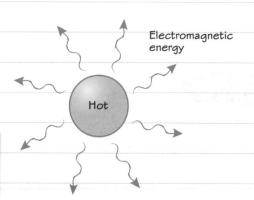

Electromagnetic energy

Hot

Now try this

Describe how a vacuum flask insulates its contents, making reference to the three processes of heat transfer.

Linear expansivity and phases of matter

Linear expansivity describes the tendency of a solid material to expand in size as it is heated. If heating is continued then eventually the solid will undergo a phase change to a liquid. Further heating will cause a phase change to a gas.

Coefficient of linear expansion (α)

Coefficient of linear expansion is a material property that describes the amount by which a material expands upon heating with each degree rise in temperature. The expansion occurs in all directions. Units 1/K.

Linear expansion is given by the equation:

$$\Delta L = \alpha L \Delta T$$

where ΔT is the temperature difference $T_2 - T_1$.

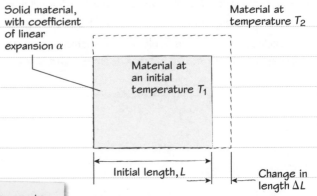

Solid material, with coefficient of linear expansion α

Material at an initial temperature T_1

Material at temperature T_2

Initial length, L

Change in length ΔL

A material undergoing thermal linear expansion.

Phase changes

The diagram shows the temperature of a material when subjected to a continuous input of heat energy. There are three distinct phases that the material adopts as its temperature increases: solid, liquid and gas.

Different phases observed when heat energy is supplied to a material at a constant rate.

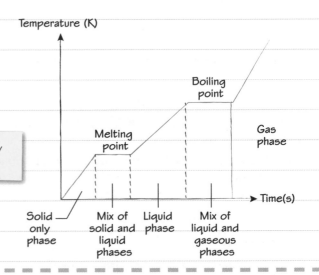

Temperature (K)

Boiling point

Melting point

Gas phase

Time(s)

Solid only phase

Mix of solid and liquid phases

Liquid phase

Mix of liquid and gaseous phases

Now try this

A steel bolt is 100 mm long at 20 °C. Given that it has a coefficient of linear thermal expansion of 12.8×10^{-6} 1/K, find the overall length of the bolt at 136 °C.

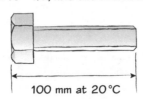

100 mm at 20 °C

Don't forget to add the extension you calculate to the original length of the bolt when finding overall length.

Specific heat capacity, latent and sensible heat

The relationship between the amount of heat energy supplied to a material and its temperature is dependent on its phase, specific heat capacity and latent heat.

Sensible heat

Any heat energy that causes a rise in the temperature of a material is known as **sensible heat**.

Specific heat capacity (c)

Specific heat capacity is the amount of heat energy required to raise the temperature of 1 kg of a material by 1 K.

The different phases of a material have different values of specific heat capacity (c_{solid}, c_{liquid} and c_{gas}).
Unit J/kg K.

Latent heat

During a phase change, heat energy is absorbed by the material and no increase in temperature is observed. This absorbed energy is known as **latent heat**.

Latent heat of fusion (L_f)

Latent heat of fusion is the amount of heat energy required for 1 kg of a material to change phase from solid to liquid.

Unit $J\,kg^{-1}$.

Latent heat of vaporisation (L_v)

Latent heat of vaporisation is the amount of heat energy required for 1 kg of a material to change phase from liquid to gas.

Unit $J\,kg^{-1}$.

Sensible heat transfer (Q)

Sensible heat transfer (Q) is a function of the mass of material (m), the specific heat capacity of the material (c) and the change in temperature (ΔT), given by:

$Q = mc\Delta T$

Unit J.

Latent heat transfer (Q)

Latent heat transfer (Q) is a function of the mass of material (m) and its latent heat (L), given by:

$Q = mL$

Unit J.

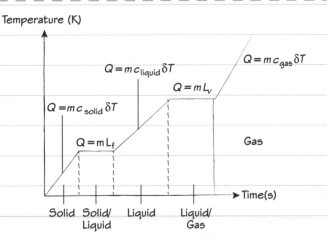

Now try this

Calculate the energy required to convert 1 kg of water at 20 °C to dry superheated steam used to drive a turbine at 146 °C.

For water: boiling point = 100 °C, L_v = 2260 kj kg^{-1}, c_{liquid} = 4.2 kj kg^{-1}K, c_{gas} = 1.8 kj kg^{-1}K.

Heat pump performance ratios

Heat pumps include refrigerators, air conditioning equipment and ground or air source heat pump heating systems.

Heat pumps

A heat pump does work in order to force heat from an area of low temperature to an area of higher temperature.

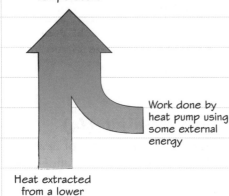

Heat delivered to a higher temperature

Work done by heat pump using some external energy

Heat extracted from a lower temperature

Cooling and heating performance

In a heat pump used for cooling (a refrigerator):

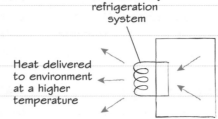

Work done by refrigeration system

Heat delivered to environment at a higher temperature

Heat extracted from interior of refrigerator at low temperature

$$\text{Refrigeration performance rate} = \frac{\text{Heat extracted}}{\text{External energy supplied}}$$

In a heat pump used for heating:

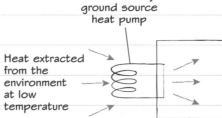

Work done by ground source heat pump

Heat extracted from the environment at low temperature

Heat delivered into a house at higher temperature

$$\text{Heat pump performance rate} = \frac{\text{Heat delivered}}{\text{External energy supplied}}$$

Worked example

A domestic ground source heat pump consumes 3 kW of electricity and can deliver 19 MJ of heat energy per hour to a house central heating system.

Determine the performance ratio of the heat pump when warming the house.

Sample response extract

From the question:

External power supplied = 3 kW = 3000 W

Heat energy delivered = 19 MJ/h = 19×10^6 J/h

Calculated quantities:

External power supplied in an hour = 3000 × 3600 = 10.8×10^6 J

$$\text{Heat pump performance ratio} = \frac{19 \times 10^6}{10.8 \times 10^6} = 1.76$$

Follow these steps:

1. Identify the parameters given in the question and convert to standard units.
2. You can use energy or power in the performance ratio calculations. In this example, use the energy used and heat delivered in an hour.
3. Calculate the performance ratio using the given formula.
4. Remember this is a performance ratio and not an efficiency. If the heat pump is working properly the performance ratio should be greater than 1.

Now try this

A domestic ground source heat pump consumes 2.5 kW of electricity and can deliver 17 MJ of heat energy per hour to a house central heating system.

Calculate the amount of heat energy per second extracted from the ground at low temperature.

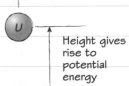
Enthalpy and entropy

Enthalpy and entropy are important thermodynamics concepts that you should be familiar with.

Internal energy (U)

Internal energy is the energy within a substance that is due to:

1 The kinetic energy of the random microscopic movement of particles inside the substance

2 The potential energy related to the forces, such as chemical bonds binding particles together.

> Internal energy does not include the potential or kinetic energy of a body on a larger scale. A stationary ball has the same internal energy as one that has been thrown up into the air.

Velocity gives rise to kinetic energy

Internal energy (U) in both cases remains the same

Height gives rise to potential energy

Stationary ball on the ground

Moving ball thrown in the air

Enthalpy (H)

Enthalpy (H) is the amount of energy contained within a thermodynamic system.

It's helpful if you think of this system as a balloon containing a fixed amount of gas.

Any heat energy entering the balloon will:

- increase the internal energy (U) of the gas or
- lead to some work being done to expand the system.

This work is the product of pressure (p) and volume (V).

Enthalpy (H) takes both these factors into account:

$H = U + pV$

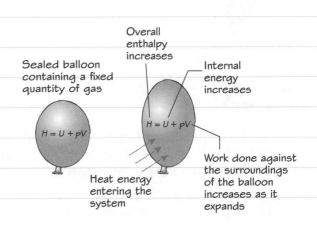

Sealed balloon containing a fixed quantity of gas

$H = U + pV$

Overall enthalpy increases

Internal energy increases

$H = U + pV$

Work done against the surroundings of the balloon increases as it expands

Heat energy entering the system

Entropy (S)

- Can be thought of as a measure of the dispersal of energy present in a thermodynamic system
- Always tends to increase
- Results in heat flow from hot to cold
- Results in fluid flow from areas of high pressure to low pressure.

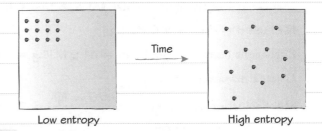

Low entropy

Time

High entropy

> An ordered system will tend towards increased disorder over time.

Now try this

Explain the difference between internal energy and enthalpy.

Thermal efficiency of heat engines

Heat engines, such as internal combustion engines and steam engines, change heat energy into mechanical work.

Heat engines

A heat engine converts some high temperature heat energy into useful mechanical work and rejects the rest to some lower temperature.

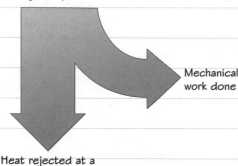

Heat energy received at high temperature

Mechanical work done

Heat rejected at a lower temperature

Thermal efficiency

The thermal efficiency of a heat engine is the ratio of useful work done to the heat energy contained in the amount of fuel used:

$$\text{Thermal efficiency} = \frac{\text{Mechanical power output (W)}}{\text{Equivalent heat energy of fuel (J/kg)} \times \text{Fuel consumption (kg/s)}}$$

Worked example

During a 20-minute test, a combustion engine did 6.96 MJ of mechanical work and used 820 g of fuel with equivalent heat energy of 41.9 MJ/kg. Calculate the thermal efficiency of the engine.

Sample response extract

From the question:

Time = 20 min = 1200 s

Mechanical work done = 6.96 MJ = 6.96×10^6 J

Fuel used = 820 g = 0.82 kg

Equivalent heat energy of fuel = 41.9 MJ/kg = 41.9×10^6 J/kg

Calculated quantities:

Mechanical power output = $\dfrac{6.96 \times 10^6}{1200}$ = 5800 W

Fuel consumption $\dfrac{0.82}{1200} = 0.68... \times 10^{-3}$ kg/s

Thermal efficiency = $\dfrac{5800}{41.9 \times 10^6 \times 0.68 \times 10^{-3}}$

= 0.204 (to 3 s.f.)

Follow these steps:
1. Identify the parameters given in the question and convert to standard units.
2. Calculate mechanical power output by dividing total work done by the duration of the test, in seconds.
3. Calculate fuel consumption by dividing the mass of fuel used by the duration of the test, in seconds.
4. Calculate thermal efficiency using the given formula.
5. Check that the solution is reasonable. You can't get more out than you put in, so an efficiency of greater than 1 is not possible!

Now try this

An internal combustion heat engine outputs mechanical power to drive a generator. The heat energy input is provided by the combustion of a fuel with an energy content of 40 MJ/kg, which is consumed at a rate of 0.005 kg/s. The mechanical power required to drive the generator is 38 kW.

Calculate the efficiency with which the heat engine provides mechanical work to the generator.

Thermodynamic process parameters

Before you can carry out any thermodynamic calculations, you must be familiar with the terms used to define them, known as process parameters.

Thermodynamic temperature (T_k)

You must use absolute or thermodynamic temperature measured in **Kelvin (K)** in thermodynamic calculations.

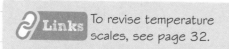 **Links** To revise temperature scales, see page 32.

Absolute zero (or 0K) is a theoretical minimum temperature when the kinetic energy of molecular vibrations within a material is zero.

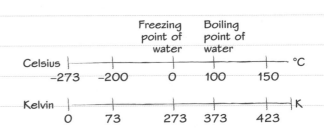

Comparison of Kelvin and Celsius temperature scales.

Volume (V)

The amount of space occupied by an object. Units m^3.

Mass (m)

The amount of matter contained within an object. Units kg.

Density (ρ)

The compactness of a material calculated as the mass (m) per unit volume (V). Units kg/m^3.

$$\rho = \frac{m}{V}$$

Pressure (P)

Always make sure you use standard SI units in calculations. The SI unit of **pressure** is the **pascal (Pa)** but several other units remain in common usage. $1\ Pa = 1\ N/m^2$.

	Pascal (Pa)	Bar (bar)	Atmosphere (atm)	Pounds per square inch (psi)
1 Pa	1	1×10^{-5}	$9.87x \times 10^{-6}$	145×10^{-6}
1 bar	1×10^5	1	0.987	14.5
1 atm	101325	1.01325	1	14.7
1 psi	6894	68.9×10^{-3}	68.0×10^{-3}	1

Comparison of units of pressure still in common usage.

Standard atmospheric pressure (P_{atm})

The pressure at the Earth's surface caused by our atmosphere is 101 325 Pa.

Gauge pressure (P_{gauge}) –

The pressure above P_{atm} that is shown on a standard pressure gauge. Units Pa.

Absolute pressure (P_{abs})

The sum of atmospheric and gauge pressure that **must be used in thermodynamic process equations.** Units Pa.

$$P_{abs} = P_{gauge} + P_{atm}$$

Pressure at any point A

P_g

Atmospheric pressure (at Earth's surface)

P_{abs}

P_{atm}

Absolute vacuum (deep space)

Now try this

A tyre pressure gauge shows a reading of 28 psi. Calculate the equivalent absolute pressure in Pa.

Gas laws

It's important for engineers to understand the behaviour of gases as they are heated or compressed. You will need to cover only the simplified behaviour of ideal gases to which these gas laws can be applied.

Boyle's law

For a fixed mass of an ideal gas at constant temperature (T), the volume (V) is inversely proportional to the pressure (P).

This can be stated as:

$PV = $ constant

or

$P_1V_1 = P_2V_2$

From Boyle's law at constant T

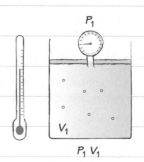

$P_1 V_1 \quad = \quad P_2 V_2$

Charles' law

For a fixed mass of an ideal gas at constant pressure (P), the volume (V) is directly proportional to the absolute temperature (T).

This can be stated as:

$\dfrac{V}{T} = $ constant

or

$\dfrac{V_1}{T_1} = \dfrac{V_2}{T_2}$

From Charles' law at constant P

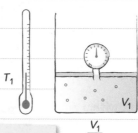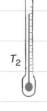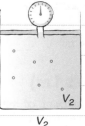

$\dfrac{V_1}{T_1} \quad = \quad \dfrac{V_2}{T_2}$

The General Gas Equation

This equation is derived by combining Boyle's and Charles' laws and takes the form:

$\dfrac{PV}{T} = $ constant

Or:

$\dfrac{P_1V_1}{T_1} = \dfrac{P_2V_2}{T_2}$

From the General Gas Equation

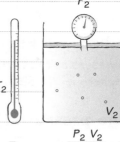

$\dfrac{P_1 V_1}{T_1} \quad = \quad \dfrac{P_2 V_2}{T_2}$

Now try this

A car tyre (with fixed volume) is inflated to an absolute pressure of 193 kPa at 20 °C. After being driven for several hours the temperature of the tyre has increased to 43 °C.

Calculate the absolute pressure in the tyre.

Remember to convert temperatures to Kelvin before doing any calculations.

Current flow

Current flow is fundamental to understanding how electricity is used by engineered components such as motors and generators. You will find the following concepts used to explain topics such as capacitors on pages 47 to 48, and DC power sources on page 57.

Atomic structure

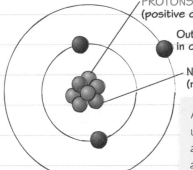

PROTONS in a nucleus (positive charge)

Outermost ELECTRONS in orbit around nucleus

NEUTRONS in a nucleus (no charge)

A solid conductor is made up of closely packed atoms with electrons available to form a 'free electron cloud'.

NUCLEUS with innermost electrons in orbit

Outermost electrons can form 'FREE ELECTRON cloud'

Static electricity

An atom consists of charged particles:

✓ The nucleus is positive and holds electrons in orbit around it.

✓ Only electrons are transferred when friction is used to charge objects.

✓ A **positively** charged object has **lost** electrons.

✓ The outer electrons are less tightly bound because they are farthest from the nucleus.

✓ In a conductor the outer electrons can drift away from the nucleus to form a 'free electron cloud'.

Metals are good **conductors** because they have many **free electrons**.

Insulators have few or no free electrons.

Current flow

Electric current, measured in amps (A), is the flow of electric charge (q) in a specified time.

The more free electrons that pass a point per second, the greater the current.

Current (I) is defined as the rate of flow of charge: $I = \frac{q}{t}$, where q is charge, in coulombs (C), and t is time measured in seconds.

Voltage applied across conductor (wire)

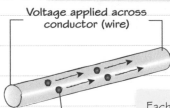

A conductor in the form of a wire in a circuit

Free electrons drift along the wire

Each electron has charge (q) of 1.6×10^{-19} C.

The free electrons drift along a conductor, such as copper wire in a circuit, when a voltage is applied across the ends of the conductor.

The direction of conventional current was decided by early experimenters before the role of electrons in current flow was understood.

A current of 2 A flows through a point in a circuit. How much charge passes the point in 3 minutes?

$I = \frac{q}{t}$

Substitute: $I = 2$ A, $t = 3 \times 60$ s

$2 = \frac{q}{180}$

therefore $q = 2 \times 180$ C = 360 C

Voltage source such as a battery

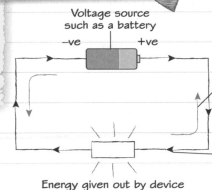

−ve   +ve

Electrons flow from the −ve terminal towards the +ve terminal

Direction of conventional current flow, from +ve terminal to −ve terminal

Energy given out by device such as a bulb or resistor

Conventional current flow

Although electrons drift from the negative to the positive terminal, the conventional **direction of current** is taken as being **from the +ve terminal towards the −ve terminal**.

1 A charge of 3 µC flows through an LED in 2 ms. Calculate the current.
2 A total charge of 4320 C passes a point in a circuit in which a current of 25 mA flows. Find the number of days this takes.

You need to be careful with the units given in question 1.

For question 2, remember that time is usually calculated in seconds, so you will need to convert this to days.

Coulomb's law and electrostatic force

You can use Coulomb's law to calculate the force between two small charged bodies.

Two equal and like charges

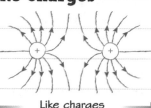

Like charges repel; the force (F) is positive.

$F \propto q_1 q_2$

A single charge (q)

The closer together the force lines are, the stronger the field.

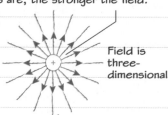

Field is three-dimensional

Field strength $\alpha \frac{1}{r^2}$, where r is the distance from the point of charge.

Two unequal and opposite charges

Higher density of force lines = more charge

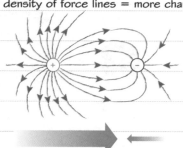

Unlike charges attract; the force (F) is negative.

$F \propto q_1 q_2$

Worked example

A positive charge of 1.68 µC and a negative charge of 4.1 µC are separated by 0.99 m. Calculate the force between the two charges.

Sample response extract

$q_1 = +1.68 \times 10^{-6}$ C, $q_2 = -4.1 \times 10^{-6}$ C,
$\varepsilon_0 = 8.85 \times 10^{-12}$ F/m, $r = 0.99$ m, $F = ?$ N

$$F = \frac{q_1 q_2}{4\pi\varepsilon_0 r^2}$$

$$F = \frac{(1.68 \times 10^{-6})(-4.1 \times 10^{-6})}{4\pi(8.85 \times 10^{-12}) \times 0.99^2}$$

$$= \frac{-6.888 \times 10^{-12}}{1.112 \times 10^{-10} \times 0.9801}$$

$$F = 63.193 \times 10^{-3} = 63.19 \times 10^{-3} \text{ N (to 2 d.p.)}$$

Coulomb's law

1 $F = \frac{q_1 q_2}{r^2}$ — Charge, in coulombs (C)

Force F, in newtons (N), between small charges

Distance (m) between the charges

2 The constant of proportionality is Coulomb's constant $(k) = \frac{1}{4\pi\varepsilon_0}$, where ε_0 is the permittivity of free space.

Don't round values during the intermediate steps of the calculation.

$F = \frac{q_1 q_2}{4\pi\varepsilon_0 r^2}$ and $\varepsilon_0 = 8.85 \times 10^{-12}$ F/m are both on the formulae sheet, so there is no need to learn them.

Permittivity of free space – uniform fields

Two parallel plates at different charge potentials have a potential voltage gradient between them.

🔗 **Links** Energy is needed to create a flow of charge between two points in a circuit. This is called potential difference or voltage. You will find further details about voltage and field strength on page 45.

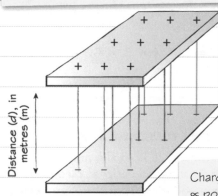

Distance (d), in metres (m)

Uniform electric field between plates

Charge on plates ∝ potential voltage difference.

Now try this

1   A negative charge of -3.0×10^{-5} C and a positive charge of 8.0×10^{-5} C are separated by 0.20 m. Calculate the force between the two charges.

2   The force between two identical charges separated by 20 mm is equal to 30 N. Find the magnitude of the two charges.

Hint: watch out for the signs of the charge.

Hint: 'identical charges' means that $q_1 = q_2$.

Resistance, conductance and temperature

Resistors are widely used in electronics and you will find details on page 44. Here we look at the factors affecting the resistance of materials and fluids.

Conductivity
- ✓ often used in relation to liquids
- ✓ constant for a material
- ✓ depends on how many free electrons are available to move
- ✓ depends on how easily the free electrons can move
- ✓ symbol sigma (σ)
- ✓ units siemens per metre (S/m)
- ✓ equal to 1/resistance or $\sigma = \dfrac{1}{\rho}$
- ✓ the higher the conductivity, the lower the resistance

Resistivity
- ✓ often used in relation to solids
- ✓ constant for a material
- ✓ depends on how many free electrons are available to move
- ✓ depends on how easily the free electrons can move
- ✓ symbol rho (ρ)
- ✓ units ohm metres (Ωm).

Resistance

Depends on its **dimensions** and the **material** it is made of.

- has symbol R
- units ohms (Ω).

Resistivity, in Ωm

Resistance, in Ω — $R = \dfrac{\rho l}{A}$ — Length, in m

Cross-sectional area, in m²

Conductance

The reciprocal of resistance is conductance (G), measured in siemens (S), i.e. $G = \dfrac{1}{R}$.

Temperature coefficient of resistance

Change in resistance (Ω)

Temperature coefficient of resistance (°C⁻¹ or K⁻¹)

$\dfrac{\Delta R}{R_o} = \alpha \Delta T$

Change in temperature, in °C or K

Original resistance (Ω)

Worked example

The resistivity of aluminium is 2.6×10^{-8} Ωm. Find the length of aluminium wire, of diameter 2.5 mm, that has resistance of 0.3 Ω.

Sample response extract

$R = 0.3$ Ω, $\rho = 2.6 \times 10^{-8}$ Ωm, $A = \pi\left(\dfrac{2.5}{2}\right)^2$ mm²

or $\pi\left(\dfrac{0.0025}{2}\right)^2$ m²

$R = \dfrac{\rho l}{A}$

Substituting values:

$0.3 = \dfrac{2.6 \times 10^{-8} \times l}{\pi\left(\frac{0.0025}{2}\right)^2}$

$l = \dfrac{\pi\left(\frac{0.0025}{2}\right)^2 \times 0.3}{2.6 \times 10^{-8}}$

$= 57$ m (to 2 s.f.)

Free electron carrying charge through lattice of the conductor

Particles in lattice – only slight oscillations

Higher temperature – greater oscillations

Conductor at lower temperature

Longer path travelled by free electron – reduced charge flow

Conductor at higher temperature

Now try this

1. A coil has a 50 Ω resistance at 50 °C and a 55 Ω resistance at 80 °C. Find its resistance temperature coefficient.

2. What is the conductance of a resistor having a resistance of 100 Ω?

The increased oscillations of particles in a heated conductor's lattice reduce the flow of free electrons by increasing the number of collisions. This increases the resistance.

Types of resistor

You need to understand the characteristics of different resistors in order to use them effectively.

Uses of fixed resistors

1 Prevent damage to electrical components

2 Control time delays in a circuit when paired with a capacitor

3 Split voltage around a circuit

Variable resistors

☑ Potentiometers: may be used either as a voltage divider, or to vary current flow.

☑ Presets: often included in circuits where the exact resistance is not known until the circuit is constructed.

Make sure that you know how a potentiometer would be wired, both as a voltage divider and as a variable resistance.

Special types

☑ Thermistors: change resistance as their temperature changes. They are used in temperature detecting circuits such as electronic thermometers.

☑ Light-dependent resistors (LDRs) change their resistance as light levels change and are used in light detection circuits such as controlling street lights.

The resistance of a thermistor usually falls as the temperature increases. The resistance of an LDR falls as light levels increase.

Worked example

A radio set contains both pre-set resistors and a potentiometer.

(a) State one use of each in a radio.

(b) Explain how often and why each would be adjusted.

Sample response extract

(a) The radio receiver circuits use a pre-set resistor. The volume is controlled with a potentiometer.

(b) The pre-set resistor is adjusted once, in the factory, to make the radio receiver work. The potentiometer is adjusted by the user many times.

Potentiometers also come in the form of a straight track called **sliders**. They are found, for example, on audio mixing decks to control volume.

Preferred resistor values

Resistors are manufactured in specific ranges to reduce the number of different values required.

The value of a resistor is shown in one of two ways:

- by colour bands; usually four or five bands
- by 'multiplication factor', where R = ×1, K = ×1000 and M = ×1000000.

The **resistor tolerance** is the amount by which the resistance of a resistor may vary from its stated value. If coloured bands are used, the tolerance is the last band, separated from the other bands by a space.

If the multiplication factor is used, then tolerance is given in the form of a letter; e.g. 'J' indicates a tolerance of ± 5%.

Now try this

1 Explain why a carbon composition resistor is unsuitable for use in a high specification audio system.

2 What is the resistance value and tolerance of a resistor specified in a circuit diagram as 5 MΩJ?

Field strength

On page 42 you used Coulomb's law to calculate the force between two small charged bodies. Here you will calculate the force on a small charged body in an electric field. There are practical aspects of this on page 60 (emf and field force), and page 67 (Faraday's laws – induced emf).

Field strength

✓ The field strength increases closer to the source charge. (You can see that the equipotential lines are closer together nearer to the source charge.)

$$E \propto \frac{1}{d^2}$$

✓ If the distance of the test charge from the source charge is doubled, then the field strength will decrease by a factor of 4 (2^2).

✓ The field strength is independent of the size of the test charge.

✓ Field strength is a vector; it has magnitude and direction.

✓ The test charge carries a quantity of charge (q).

✓ The +ve test charge will experience a force (F) of attraction towards the −ve source charge.

✓ The electric field strength (E) is defined as the force per charge on the test charge or:

$$E = \frac{F}{q}$$

Non-uniform electric field

Here is a (-ve) source charge with a (+ve) test charge placed at some distance from it to measure the field strength at that point.

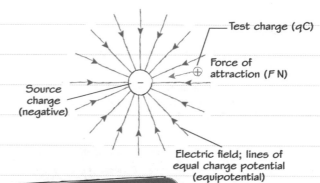

Test charge (qC)

Force of attraction (F N)

Source charge (negative)

Electric field; lines of equal charge potential (equipotential)

Worked example

A test charge of 1×10^{-6} C experiences a repulsion force of 0.40 N when placed within the electric field of a positively charged source of 2×10^{-4} C. Find the electric field strength at that point.

Sample response extract

$$E = \frac{F}{q} = \frac{0.4}{1 \times 10^{-6}} = 4.0 \times 10^5 \text{ N/C}$$

Worked example

A food preservation process uses pulsed electric fields (PEF) between parallel conducting plates. The uniform field strength is 40 kV/cm. Calculate the voltage applied to the plates if they are 0.3 m apart.

Sample response extract

$$E = \frac{V}{d}$$

$$40 \text{ kV/cm} = 40 \times 10^3 \times 100$$

$$= 4.0 \times 10^6 \text{ V/m}$$

Substituting: $4 \times 10^6 = \dfrac{V}{0.3}$

$$V = 4 \times 10^6 \times 0.3 \text{ V}$$

$$= 1.2 \times 10^6 \text{ V or } 1.2 \text{ MV}$$

Uniform electric field

This is a special case when considering electric field strength. Instead of an inverse square relationship, the field strength is constant between the two plates (but not at the edges!).

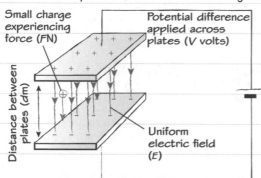

Small charge experiencing force (F N)

Potential difference applied across plates (V volts)

Distance between plates (d m)

Uniform electric field (E)

The field strength is constant, therefore the force on a small charged object is constant.

Field strength (V/m)

$$E = \frac{V}{d}$$

Potential difference applied to the plates (V)

Distance between the plates (m)

By convention, the direction of the field is from the positive to the negative plate.

Now try this

1 A test charge of 2×10^{-5} C experiences a repulsion force of 0.20 N when placed within the electric field of a positively charged source of 3×10^{-4} C. Determine the electric field strength at that point.

2 Find the electric field strength between a pair of parallel conducting plates, positioned 4 cm apart, when the potential difference between them is 300 V.

Had a look ☐   Nearly there ☐   Nailed it! ☐

Capacitance

There are three factors that effect the amount of charge stored on parallel plates. You will find details of how these are used in electrical capacitors on pages 47 and 48.

Charge on parallel plates

Amount of charge stored depends on:
- ✓ distance between the plates
- ✓ area of the plates
- ✓ material between the plates.

Units of capacitance

1F capacitor, charged by 1V carries a charge of 1C.

This is a large amount of charge; usually capacitors have values in the microfarad (10^{-6}) or picofarad (10^{-12}) range.

Note that flux density (quantity of charge per unit area ($D = Q/A$)) remains the same.

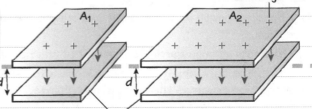

Reducing d increases field strength and therefore capacitance.

$C \propto \frac{1}{d}$

Increased field strength

Increased quantity of charge

Increasing A increases charge on the plates and therefore capacitance.

$C \propto A$

Permittivity

- Symbol for absolute permittivity is ε (Greek letter epsilon).
- Units of permittivity are farads/metre or F/m.
- The permittivity depends on the material between the plates, called the **dielectric**.
- ε is usually written $\varepsilon_r \varepsilon_0$
- ε_r is the relative permittivity of the dielectric used. The values are relative to a vacuum, which has a value of 1.
- ε_0 is the permittivity of free space (8.85×10^{-12} F/m).
- Typical values of ε_r are 2.5 for paper, 6 for mica and 80 for water. Air has a value very close to 1.

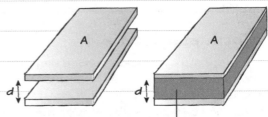

Dielectric (an insulator) between the plates increases capacitance. $C \propto \varepsilon$ where ε is the permittivity of the dielectric.

Calculating capacitance

Absolute permittivity ε is the constant of proportionality in the relationship $C \propto \frac{A}{d}$ such that $C = \varepsilon \frac{A}{d}$. This is more usually written as $C = \varepsilon_r \varepsilon_0 \frac{A}{d}$.

- The value of ε_0 (8.85×10^{-12} F/m) is provided on the formulae and constants sheet.
- The formula will be on the formula sheet in the form $C = \varepsilon \frac{A}{d}$. Remember to substitute ε with $\varepsilon_r \varepsilon_0$ so that you use the form $C = \varepsilon_r \varepsilon_0 \frac{A}{d}$.

Worked example

A parallel plate capacitor in a circuit has plates of area 0.005 m^2 and 1 mm apart, with a polymer dielectric of relative permittivity 3.5. Calculate the capacitance.

Sample response extract

$C = \varepsilon_r \varepsilon_0 \frac{A}{d}$ F/m, substituting:

$C = \frac{3.5 \times 8.85 \times 10^{-12} \times 0.005}{1 \times 10^{-3}} = 1.54875 \times 10^{-10}$ F

$= 154.87 \times 10^{-12}$ F or 154.87 pF

Now try this

A parallel plate capacitor in a circuit has plates of area 0.01 m^2 and 1 mm apart, with an electrolytic-based dielectric of relative permittivity $\varepsilon_r = 2$. Calculate the capacitance.

Capacitors – non-polarised

Capacitors find many uses in electronic circuits and some of these are covered on pages 53 and 54. Here and on page 48 you will explore the three main types of capacitor and their construction.

Types of capacitor

Three main groups:

Polar capacitors

Tantalum

Non-polar capacitors

Electrolytic

Ceramic

Polyester

Variable capacitors

Electrolytic

Capacitor construction – non-polarised

May be used with AC or DC voltages
- fixed value
- film.

✓ Waxed paper or polymer dielectric between metal foil.

✓ Named after their dielectric, for example, paper, PP or PTFE.

✓ Typical relative permittivity: 2.5 for polymer and 4 for paper.

✓ Typical capacitance values: between 0.5 µF and 50 pF for polymer, and 10 µF and 10 nF for paper.

✓ Working voltages are up to about 600 V for paper and 400 V for polymer.

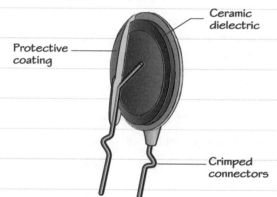

Insulating paper

Metal films

Ceramic

Ceramic dielectric

Protective coating

Crimped connectors

✓ Two types: class 1 and class 2, each with different characteristics.

✓ Simplest construction is a plate of ceramic coated with silver on each side. Often with crimped wires to keep it clear of PCB-induced movement or stress.

✓ Typical relative permittivity: Class 1: 20 to 40, class 2: 200–14 000.

✓ Typical capacitance values: 1 µF to 5 pF.

✓ Rated voltages up to 500 V.

Variable value

✓ The simplest form has an air dielectric with movable plates.

✓ Relative permittivity of 1.

✓ Typical capacitance values are 1 nF to 100 pF.

✓ Working voltages 10 V to 2 kV.

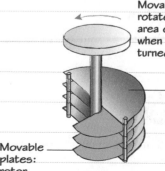

Movable plates rotate to change area of plate overlap when the control is turned.

Fixed plates: stator

Movable plates: rotor

Now try this

Explain why the dielectric material in a capacitor must be an insulator.

Capacitors – polarised

Capacitors find many uses in electronic circuits and some of these are covered on pages 53 and 54. Following on from the construction of non-polarised capacitors on page 47, here you will explore the construction of polarised capacitors and the calculation of dielectric field strength.

Capacitor construction – polarised

DC voltage only; correct terminal connection is required to avoid breaking down the insulating oxide layer.

Supercapacitors

Supercapacitors bridge the gap between electrolytic capacitor and rechargeable battery.

Electrolytic

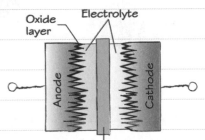

Oxide layer   Electrolyte

Anode   Cathode

Paper saturated with electrolyte

✓ Relies on chemical action for operation; oxide forms on +ve plate when voltage applied. This acts as dielectric.

✓ Historically named after cathode material, for example, aluminium, tantalum or niobium.

✓ May be radial or axial.

✓ Relative permittivity: 10 to 40.

✓ Typical capacitance values: 1F to 1μF.

✓ Working voltages 10–600 V.

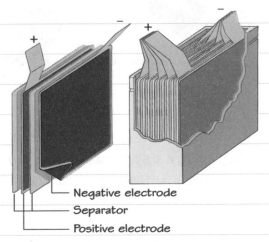

Negative electrode
Separator
Positive electrode

✓ Very high capacitance values but lower voltage limits than other capacitors.

✓ Used where multiple charge/recharge required, such as regenerative braking or CPU cache memory back-up in computers.

Dielectric strength

The electric field strength is defined as $E = \dfrac{V}{d}$, where E is the electric field, V is the potential difference (in volts) and d is the distance between the plates (in metres).

An applied potential difference greater than rated capacitor voltage can strip atoms of electrons and make the dielectric conduct. Charge is no longer maintained on the plates.

Typical dielectric strengths:

- dry air 3×10^6 V/m
- ceramics 10×10^6 V/m
- polymers 20×10^6 V/m
- mica 40×10^6 V/m

Worked example

The dielectric strength (E) of a polymer is 20×10^6 V/m.

Calculate the maximum potential difference that can be applied to a parallel plate capacitor with polymer dielectric thickness 0.2 mm.

Sample response extract

$$E = \frac{V}{d}$$

$$V_{max} = E_{max}d$$

Substitute in $V_{max} = 20 \times 10^6 \times 2 \times 10^{-4}$ V/m

$$= 4000\,V$$

$$= 4\,kV$$

Now try this

1   Explain why the terminals on polarised capacitors must have polarity markings. Provide an example of how this is shown.

2   An electrolytic capacitor has a tantalum oxide dielectric of thickness 1 μm and dielectric strength 5×10^6 V/m. What is the maximum voltage that can be used with this capacitor?

Ohm's law, power and efficiency 1

Ohm's law applies to OHMIC CONDUCTORS, whether in a DC or AC circuit. Providing the temperature remains constant, their resistance doesn't change over a wide range of currents.

If you are given two of the values, you can calculate the unknown third value.

1 $V = IR$ **2** $I = \dfrac{V}{R}$ **3** $R = \dfrac{V}{I}$

Where V = potential difference measured in volts (V), I = current measured in amps (A), and R = resistance measured in ohms (Ω).

In graphical form

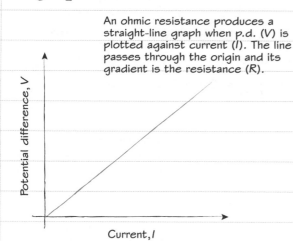

An ohmic resistance produces a straight-line graph when p.d. (V) is plotted against current (I). The line passes through the origin and its gradient is the resistance (R).

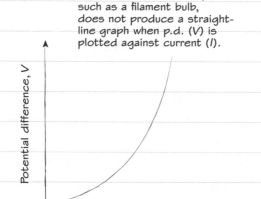

A non-ohmic resistance, such as a filament bulb, does not produce a straight-line graph when p.d. (V) is plotted against current (I).

Worked example

A thermistor of resistance $10\,k\Omega$ at room temperature is connected across a 12 V source.

(a) Calculate the resulting current.

(b) Sketch a graph of potential difference against current for a thermistor.

(c) Explain whether thermistors obey Ohm's law.

Sample response extract

(a) $V = 12\,V$, $I = ?$, $R = 10 \times 10^3\,\Omega$

     $V = IR$ $12 = 10 \times 10^3\,I$

     $I = \dfrac{12}{10 \times 10^3} = 1.2 \times 10^{-3}\,A$ or $1.2\,mA$

(b)

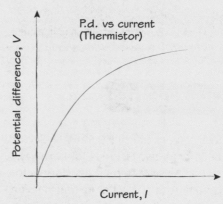

P.d. vs current (Thermistor)

(c) Although it goes through the origin, the resistance (gradient: $\dfrac{V}{I}$) of the thermistor shown in the sketch of p.d. against current varies as the current changes: $R \neq \dfrac{V}{I}$, therefore it does not obey Ohm's law.

Now try this

1. Sketch the graph of current against voltage for a diode and use it to explain why a diode is a non-ohmic device. Your revision on page 56 may help with this.

2. A power tool is supplied with 110 V and has a power rating of 1.5 kW. Calculate the current drawn by the motor.

Ohm's law, power and efficiency 2

Power and energy are very important to engineers when designing electric/electronic products. Whether increasing battery life of a mobile phone by reducing power demands, or reducing the noise from a computer by reducing the need for cooling air, the power equations and efficiency calculations are useful in optimising engineering solutions at the design stage.

Variations on the power equation

Energy transferred to a circuit component is the product of the number of coulombs per second (current in amps, A) and the energy per coulomb (potential difference in volts, V). i.e. Power in watts = VI. But Ohm's law tells us $I = \dfrac{V}{R}$ so:

✓ $P = V \times \dfrac{V}{R}$ or $P = \dfrac{V^2}{R}$

Similarly $V = IR$ so:

✓ $P = IR \times I$ or $P = I^2R$

Engineers refer to wasted energy lost in power cables as I^2R losses.

Worked example

A 300 Ω resistor has a power rating of 4 W. Find the maximum current that can be drawn through the resistor without exceeding the power rating.

$R = 300\ \Omega,\ P = 4\ W,\ I = ?\ A;\quad P = I^2R$

Substituting: $4 = I^2 \times 300,\quad I^2 = \dfrac{4}{300}$

$I = \sqrt{\dfrac{4}{300}} = 0.116\ A$ or 116 mA

By transmitting power at higher voltage through the same resistance, the current is correspondingly lower (think Ohm's law: $R = \dfrac{V}{I}$). Therefore I is greatly reduced, which reduces the wasted energy.

Efficiency

Usually expressed as a percentage:

$E = \left(\dfrac{\text{Power out}}{\text{Power in}}\right) \times 100\%$

Here you are looking at power, which is the energy consumed per unit time and is measured in joules per second or watts (W).

Electrical energy from power station B: 2.5 MW

Electrical energy received at factory 2.0 MW

I^2R heat losses: 0.5 MW

Sankey diagram.

Worked example

Two power stations transmit 2.5 MW of electrical power at 25 kV for two factories 100 km away. Power station A uses a step-up transformer to raise the voltage to 400 kV before transmission, but power station B does not. In each case the resistance of the transmission line is 50 Ω.

Calculate:

(a) the power lost in transmission lines for each power station

(b) the percentage efficiency of each.

Sample response extract

Current generated by station A: $P = IV$,

$2.5 \times 10^6 = I \times 400 \times 10^3,\ I = \dfrac{2.5 \times 10^6}{400 \times 10^3} = 6.25\ A$

a) Power lost due to heating power line = I^2R

   $= 6.25^2 \times 50 = 1.95\ kW$

b) Efficiency = $\dfrac{2.5 \times 10^6 - 1.95 \times 10^3}{2.5 \times 10^6} \times 100$

   $= 99.92\%$ (to 2 d.p.)

Current generated by station B: $P = IV$,

$2.5 \times 10^6 = I \times 25 \times 10^3,\ I = \dfrac{2.5 \times 10^6}{25 \times 10^3} = 100\ A$

(a) Power lost due to heating power line = I^2R

   $= 100^2 \times 50 = 0.50\ MW$

(b) Efficiency $= ((2.5 - 0.5)/2.5) \times 100 = 80\%$

Now try this

1   If the voltage across a circuit is quadrupled, explain how the current through the circuit changes. Assume the power remains constant.

2   Calculate the efficiency of a power supply that produces a 0.6 W output from a 5 V supply and draws 0.2 A.

Kirchoff's voltage and current laws

You can use Ohm's law to analyse a simple circuit but you also need Kirchoff's laws for networks of resistors.

Kirchoff's voltage law

When resistors are connected in series the sum of the voltage drop (or potential differences PD) across each one is equal to the total supply voltage.

So where three resistors are in series then:

$V = V_1 + V_2 + V_3$

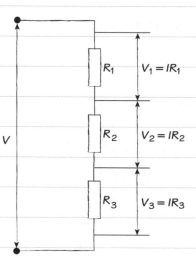

Combining Kirchoff's and Ohm's law

When you combine Kirchoff's voltage law with Ohm's law you get:

$V = V_1 + V_2 + V_3 = 1R_1 + 1R_2 + 1R_3$

This is sometimes written in terms of the sum of the potential differences (or voltage drops) across each resistor: $\Sigma PD = \Sigma 1R$.

Kirchoff's current law

At any junction in an electrical circuit, the total current flowing towards the junction is equal to the total current flowing away from it.

When I flows towards the junction and I_1, I_2 and I_3 flow away from it then $I = I_1 + I_2 + I_3$.

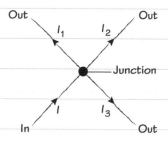

The diagram shows a simple series circuit containing three resistors.
Calculate the value of resistor R.

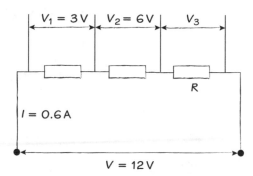

Capacitors – charging and energy

Capacitors may be found in almost every electrical product. Just a few of their uses are to smooth power supplies, filter wanted or unwanted signals or provide energy storage. Here you will calculate the charge and the energy stored by a capacitor. On pages 53 and 54 you will revise how they can be used in circuits.

Charging a capacitor

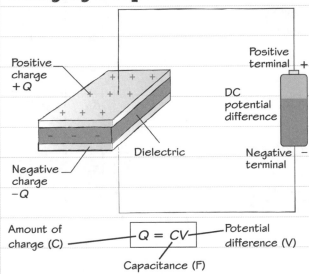

Positive charge +Q

Positive terminal +

DC potential difference

Dielectric

Negative terminal −

Negative charge −Q

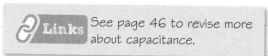

Amount of charge (C) —— $Q = CV$ —— Potential difference (V)

Capacitance (F)

> **Links** See page 46 to revise more about capacitance.

Worked example

A DC 24 V circuit contains a 3 µF capacitor. Calculate the charge stored by the capacitor.

Sample response extract

$Q = ?$, $C = 3 \times 10^{-6}$ C, $V = 24$ V

$Q = CV = 3 \times 10^{-6} \times 24$

$\qquad = 72 \times 10^{-6}$ C or 72 µC

The symbol for coulomb is an upright C, whereas the variable capacitance is an italic C. The algebraic sum of charge on the capacitor plates is 0 because one plate is +Q and the other is −Q. Q refers to the **magnitude** of charge stored on each plate. The fully charged 3 µF capacitor will have +72 µC on the positive plate and −72 µC on the negative plate.

Energy stored in a capacitor

✓ The energy stored $E = \frac{1}{2} CV^2$

✓ The energy, in joules, stored in a capacitor is equal to the work done in charging the capacitor from 0 V.

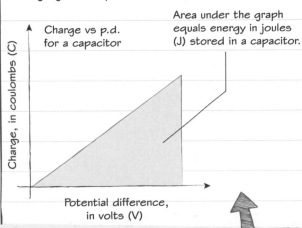

Charge vs p.d. for a capacitor

Area under the graph equals energy in joules (J) stored in a capacitor.

Charge, in coulombs (C)

Potential difference, in volts (V)

A capacitor is never connected directly to a voltage source. There will always be a resistor to control the rate at which the capacitor is charged – this protects both the capacitor and the voltage source.

Worked example

6.25×10^{-3} J of energy needs to be stored on a capacitor that is supplied with 200 V. Calculate the capacitance value required.

Sample response extract

$E = 6.25 \times 10^{-3}$ J, $C = ?$, $V = 200$ V

$E = \frac{1}{2} CV^2$

$6.25 \times 10^{-3} = \frac{1}{2} \times C \times 200^2$

$C = \dfrac{2 \times 6.25 \times 210^{-3}}{200^2}$

$\qquad = 0.31$ µF

> **Links** Revise more about this on page 55.

Now try this

A 3 mF capacitor is connected to a 6 V DC power supply. How much charge can be stored by the capacitor?

Capacitors – networks

Here you will revise the different results from connecting capacitors in series and in parallel.

Capacitors in parallel and series

Parallel

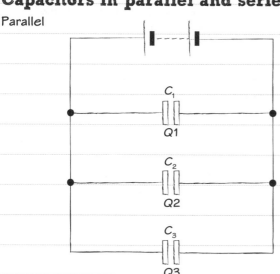

$C_T = C_1 + C_2 + C_3$

Series

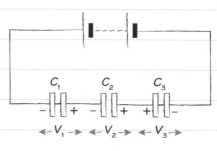

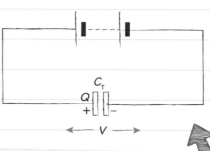

$\frac{1}{C_T} = \frac{1}{C_1} + \frac{1}{C_2} + \frac{1}{C_3}$

The potential difference applied will equal the sum of the potential differences across each capacitor: $V = V_1 + V_2 + V_3$

Find the total capacitance between points X and Y in the circuit shown.

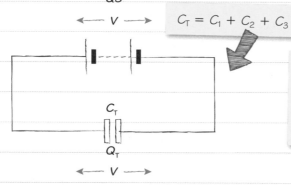

First consider the two $3\,\mu F$ capacitors; they are in parallel and so can be replaced by one capacitor of value $6\,\mu F$ ($C_T = C_1 + C_2$).

Next, consider the two capacitors in series: $6\,\mu F$ and $2\,\mu F$: $C_T = \frac{C_1 C_2}{C_1 + C_2}$.

Remember 'product over sum' works only for two capacitors. For three or more, you need to calculate the reciprocals individually and then take the reciprocal of the answer to obtain C_T:

i.e. $\frac{1}{C_T} = \frac{1}{C_1} + \frac{1}{C_2} + \frac{1}{C_3}$ … etc.

For parallel capacitors: $C_T = C_1 + C_2$
$= 3 + 3 = 6\,\mu F$

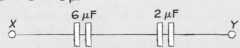

The two capacitors are in series: $\frac{1}{C_T} = \frac{1}{C_1} + \frac{1}{C_2}$

$C_T = \frac{C_1 C_2}{C_1 + C_2} = \frac{6 \times 2}{6 + 2} = \frac{12}{8} = 1.50\,\mu F$ (to 2 d.p.)

Now try this

1 When using identical capacitors, state which has the largest capacitance.

   (a) three capacitors in series

   (b) three capacitors in parallel

2 Two capacitors with values of $35\,\mu F$ and $76\,\mu F$ are connected in series. What is the total capacitance of this arrangement?

Capacitors in circuits – the time constant

Capacitors are charged and discharged through a resistor, which controls the flow of current and therefore the time taken for the capacitor to charge and discharge. On the next page you will revise how to analyse the RC transients during the exponential-based charging and discharge cycle.

Capacitor charging

The current flow decays. The potential difference across the plates increases as the charge accumulates.

Capacitor discharging

The current flow, potential difference across the plates and the charge all follow the same type of decay curve during the discharge.

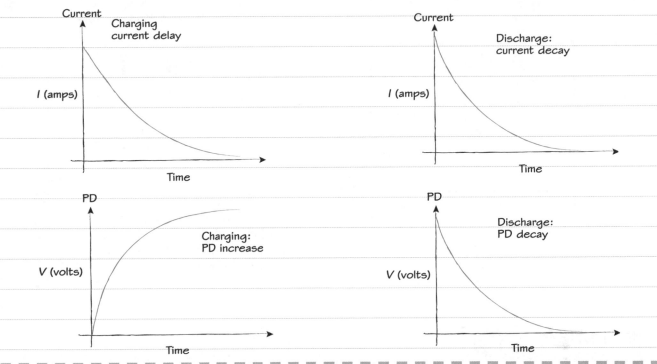

RC transient period

The generally accepted time for a capacitor to be fully charged/discharged is taken as 5 time constants. The RC **transient period** is taken to be 0 > time constants < 5.

RC transients

The time constant (Greek letter 'tau' τ), in seconds, for the capacitor to charge/discharge depends on the capacitance of the capacitor (C, in farads) and the resistance of the resistor (R, in ohms).

It is given by the product of the two values:

$\tau = RC$

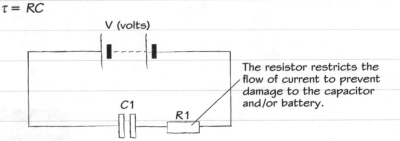

The resistor restricts the flow of current to prevent damage to the capacitor and/or battery.

Now try this

A 2200 µF capacitor is charged to 12 V through a 100 kΩ resistor. Calculate

(a) the maximum charge stored by the capacitor

(b) the time constant of the circuit.

Capacitors in circuits – *RC* transients

You need to know how to analyse the *RC* transients during the exponential-based charging and discharge cycles of capacitors.

Voltage change on capacitor charge/discharge

The value of the voltage across the capacitor (V_c) at any instant in time during the charge period is given by $V_c = V_s(1 - e^{\frac{-t}{\tau}})$.

The equivalent discharge voltage is given by $V_c = V_s e^{\frac{-t}{\tau}}$, where V_s is the supply voltage and t is the elapsed time since the application/removal, respectively, of the supply voltage.

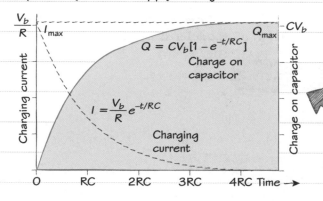

Worked example

An extrusion machine in a plastic bottle manufacturing factory includes an *RC* network with a time constant of 0.56 s. The value of the capacitor is 3.9 μF. Calculate the value of the resistor.

Sample response extract

$\tau = 0.56$, $R = ?$, $C = 3.9\,\mu F$

$\tau = RC$

$0.56 = 3.9 \times 10^{-6} R$

$R = \dfrac{0.56}{3.9 \times 10^{-6}} = 144\,k\Omega$

If these expressions are required for the exam, they will be provided on the sheet. There is no need to learn them.

Make sure you know the laws of logs and indices, which can be revised on pages 1 and 2.

Worked example

A single-phase squirrel cage motor requires a secondary winding, in series with a 'start capacitor', to generate sufficient torque to start the motor turning. A typical start capacitor has a capacitance of 100 μF and a supply voltage of 250 V. What is the discharge voltage after 2 time constants?

$V_c = ?\,V$, $V_s = 250\,V$, $t = 2\tau$, $\tau = ?$ Ignore the capacitance because you don't need to calculate the time constant, τ.

$V_c = V_s e - t/\tau$

$V_c = 250 e^{-2\tau/\tau}$

$V_c = 250 e^{-2} = 250 \times 0.1353$

$V_c = 33.83\,V$ (to 2 d.p.)

Don't forget to change the sign of the power 2 to a minus.

Use the 'ln' button on your calculator, together with the 'shift' key, to access the e^x function.

Now try this

1. An *RC* timing circuit has a time constant of 0.56 s. The value of the resistor is 15 kΩ. Calculate the value of the capacitor.

2. An *RC* circuit has a time constant of 5 seconds. How long should be allowed for the capacitor to be considered fully charged?

Diodes – bias and applications

Here you will explore how diodes work and the many uses they have in electronic engineering.

Bias characteristics

Simple diodes allow current to flow in one direction only. Modern diodes are based on a semiconductor pn junction. The voltage/current characteristics can be amended by the addition of doping material at manufacture.

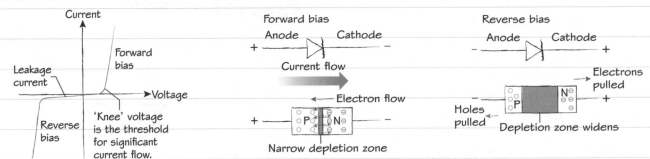

Remember: **forward** bias is **positive** terminal to **p** side of the junction. **Reverse** bias is **negative** terminal to **p** side of the junction. P (Anode) side of junction has holes. N (Cathode) side of junction has electrons.

Forward bias applications

Remember that before a diode will allow significant current to pass there must be an applied voltage to collapse the depletion region. For germanium diodes the forward voltage is 0.3 V and for silicon it is 0.7 V.

You should be able to sketch a suitable diagram to support your description of uses for a forward-biased diode.

Worked example

Explain how a diode can provide:

(a) rectification   (b) clamping function   (c) circuit protection.

Sample response extract

(a) One diode can provide half wave rectification of an AC current, but full wave rectification requires four diodes in a bridge circuit.

(b) A diode connected with reverse bias in parallel to a component protects the circuit from back EMF when the component is switched off.

(c) A series-connected diode protects the circuit if the battery is connected the wrong way round.

Reverse bias

In reverse bias:

☑ a small leakage current, although there is a high resistance

☑ avalanche breakdown or Zener effect, application of sufficient (precise) voltage causes a significant current to flow.

Typical uses for Zener reverse-biased diodes include:

① voltage regulator

② power dissipation

③ reference voltage generation.

Worked example

Describe how a Zener diode can be used to regulate voltage in a circuit.

Sample response

A Zener diode provides a controlled breakdown when reverse biased. It will try to limit the voltage drop across it to the breakdown voltage even if the current changes. The voltage drop is very nearly constant across a wide range of reverse currents.

Now try this

Describe, with the aid of a sketch, how you could achieve full wave rectification using a bridge rectifier containing four diodes.

DC power sources

Direct current (DC) power sources are able to provide the voltage and current required to power a range of electrical and electronic devices.

Cells

Cells are electrochemical devices able to generate an emf and are used as a source of electrical current.

Batteries

Batteries contain multiple cells connected together to provide higher voltage or current.

Internal resistance

The internal structure of a cell or battery has some resistance to the flow of current. This is known as internal resistance and is the reason why batteries tend to get hot when discharged rapidly.

> You can calculate the internal resistance of a battery by considering it as an additional series resistor in a circuit.

> You might have to remind yourself about Ohm's law and resistors in series to follow this worked example.

Worked example

The circuit contains a battery supplying an emf of 12 V. A current of 1.1 A flows through an external load of 10 Ω.

Calculate the internal resistance of the battery.

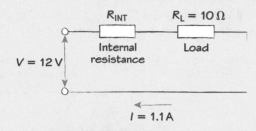

R_{INT} $R_L = 10\ \Omega$

Internal resistance   Load

$V = 12\,V$

$I = 1.1A$

Sample response extract

Calculate total resistance, R_T, using Ohm's law:

$R_T = \frac{V}{I} = \frac{12}{1.1} = 10.909\,\Omega$

For resistors connected in series:

$R_T = R_{INT} + R_L$

Rearrange to make R_{INT} the subject:

$R_{INT} = R_T - R_L$

$R_{INT} = 10.909 - 10 = 0.909\,\Omega$

Stabilised power supply

- mains powered
- produces an accurately regulated pre-set DC output
- more versatile, controllable and safer than batteries
- found in science labs and electronics workshops.

Photo-voltaic cells

- semiconductor devices that generate an emf when exposed to light
- generate small voltages
- usually connected together into large arrays to provide higher voltages
- can be mounted on roofs for domestic use.

Now try this

A battery with an emf of 12 V and internal resistance 0.5 Ω is connected to a load with resistance 9 Ω. Calculate the voltage across the battery terminals.

Resistors in series or parallel

You need to know how to analyse circuits that contain resistors connected together in series or parallel.

Resistors in series

When resistors are connected in **series** the **same current** flows throughout the circuit and the current flowing through each resistor is the same.

The total resistance (R_T) for a number (n) of resistors in series is simply:

$R_T = R_1 + R_2 \ldots + R_n$

> Resistors in **series** where current (I) is constant.

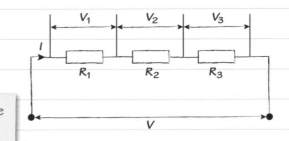

Resistors in parallel

When resistors are connected in **parallel** there is the **same voltage** across each resistor.

The total resistance (R_T) for a number (n) of resistors in parallel is:

$\dfrac{1}{R_T} = \dfrac{1}{R_1} + \dfrac{1}{R_2} \ldots + \dfrac{1}{R_n}$

> Resistors in **parallel** where voltage (V) is constant.

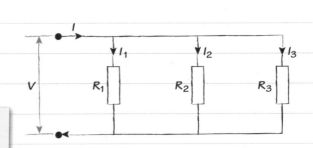

Worked example

A simple circuit contains three resistors connected in parallel, where $R_1 = 1\,k\Omega$, $R_2 = 2.2\,k\Omega$ and $R_3 = 5.7\,k\Omega$. The circuit supply voltage $V = 12\,V$.

Calculate the circuit current, I.

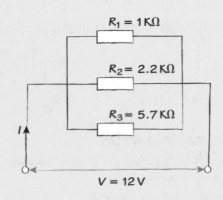

$R_1 = 1\,K\Omega$

$R_2 = 2.2\,K\Omega$

$R_3 = 5.7\,K\Omega$

$V = 12\,V$

Sample response extract

For resistors in parallel:

$\dfrac{1}{R_T} = \dfrac{1}{R_1} + \dfrac{1}{R_2} + \dfrac{1}{R_3} = \dfrac{1}{1} + \dfrac{1}{2.2} + \dfrac{1}{5.7}$

$\dfrac{1}{R_T} = \dfrac{1}{R_1} + \dfrac{1}{R_2} + \dfrac{1}{R_3} = \dfrac{1}{1} + \dfrac{1}{2.2} + \dfrac{1}{5.7}$

$R_T = \dfrac{1}{1.981} = 0.614\,k\Omega$

Apply Ohm's law:

$I = \dfrac{V}{R} = \dfrac{12}{0.614 \times 10^3} = 0.0195\,A$ (to 3 s.f.)

Follow these steps:
1. Annotate the circuit diagram.
2. Find the total resistance of the parallel resistors.
3. Use Ohm's law to find the current.

Now try this

A simple circuit contains three resistors connected in parallel. The circuit supply current $V = 12\,V$ and current $I = 25\,mA$. Two of the resistors have known values, where $R_1 = 2.2\,k\Omega$ and $R_2 = 5.3\,k\Omega$.

Calculate the value of resistor R_3.

Resistors in series and parallel combinations

It is common for electrical and electronic circuits to contain resistors connected together in series and parallel combinations. It's important that you understand the voltage and current characteristics of these networks.

Series and parallel combination

First calculate the equivalent resistance of the parallel element, using

$$\frac{1}{R_p} = \frac{1}{R_1} + \frac{1}{R_2}$$

Then use:

$R_T = R_p + R_3$ to find the total resistance of the combination.

Resistors connected in a **series and parallel** combination.

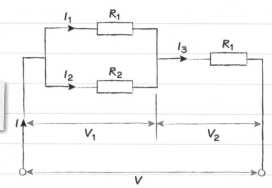

Worked example

In the circuit shown, calculate current I.

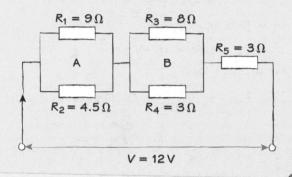

$R_1 = 9\,\Omega$ $R_3 = 8\,\Omega$

A   B $R_5 = 3\,\Omega$

$R_2 = 4.5\,\Omega$ $R_4 = 3\,\Omega$

$V = 12\,V$

You may need to remind yourself about Ohm's law before working through this example.

Sample response extract

Simplifying parallel combination A:

$$\frac{1}{R_A} = \frac{1}{R_1} + \frac{1}{R_2} = \frac{1}{9} + \frac{1}{4.5}$$

$R_A = 3\,\Omega$

Simplifying parallel combination B:

$$\frac{1}{R_B} = \frac{1}{R_3} + \frac{1}{R_4} = \frac{1}{9} + \frac{1}{3}$$

$R_B = 2.18\,\Omega$

Combining series elements of simplified circuit:

$R_T = R_A + R_B + R_5 = 3 + 2.18 + 3$

$R_T = 0.18\,\Omega$

Apply Ohm's law:

$$I = \frac{V}{R_T} = \frac{12}{8.18} = 4.47\,A \text{ (to 3 s.f.)}$$

Follow these steps:
1 Identify the parts of the circuit that can be simplified and annotate the circuit diagram.
2 Find the combined total resistance of R_1 and R_2 in parallel. Call this R_A.
3 Find the combined total resistance of R_3 and R_4 in parallel. Call this R_B.
4 Find the combined total resistance of R_A, R_B and R_5 in series.
5 Apply Ohm's law to find I.

Now try this

In the circuit shown calculate the value of R_3.

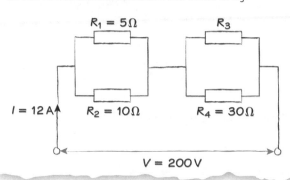

$R_1 = 5\,\Omega$ R_3

$I = 12\,A$ $R_2 = 10\,\Omega$ $R_4 = 30\,\Omega$

$V = 200\,V$

Use Ohm's law to find the equivalent total resistance. Next find the equivalent total resistances of the parallel combinations.

Substitute in $\frac{1}{R_p} = \frac{1}{R_3} + \frac{1}{R_4}$ to find the unknown value R_3.

Resistors and diodes in series

It is common for electrical and electronic circuits to contain diodes connected together in series with resistors. It's important that you understand the voltage and current characteristics of these simple circuits.

Resistors and diodes in series

When a forward biased diode is connected in series with a resistor there is a voltage drop across the diode.

For silicon diodes the voltage drop is typically 0.7 V.

From Kirchoff's voltage law, the total voltage (V_T) for a silicon diode and resistor in series is:

$V_T = 0.7 + V_2$

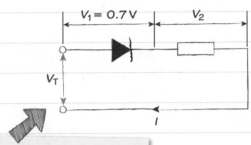

Resistor connected in **series** with a forward biased silicon diode.

A simple circuit contains a typical silicon diode and resistor connected in series. The voltage drop across the diode is 0.7 V. The supply voltage is 12 V DC and $R_1 = 3.3 \, k\Omega$.

Find the current flowing in the circuit.

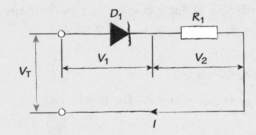

$V_T = 12 \, V$, $V_1 = 0.7 \, V$:

$V_2 = 12 - 0.7 = 11.3 \, V$

From Ohm's law:

$I = \dfrac{V_2}{R_1} = \dfrac{11.3}{3.3 \times 10^3} = 0.0034 \, A$ (to 2 s.f.)

Follow these steps:
1. The voltage drop for a silicon diode is 0.7 V.
2. From Kirchoff's voltage law: $V_T = V_1 + V_2$. Rearrange to make V_2 the subject and substitute known values to find V_2 across known resistor R_1.
3. Apply Ohm's law to find the current flowing through R_1, which will be the same throughout a series circuit.

A typical silicon diode used in a wide range of electronic circuits.

Now try this

A simple circuit contains a silicon diode and resistor connected in series. The supply voltage is 9 V DC and the circuit current is 6 mA. The voltage drop across the diode is 0.7 V.

Find the resistance of the series resistor.

Capacitors in series or parallel

You need to know how to analyse circuits that contain capacitors connected together in series or parallel.

Capacitors in series

When capacitors are connected in **series** the **charge (Q)** stored by each capacitor is the same.

The total capacitance (C_T) for a number (n) of capacitors in series is $\frac{1}{C_T} = \frac{1}{C_1} + \frac{1}{C_2} \ldots - \frac{1}{C_n}$.

Note that capacitors in series and parallel are treated differently from resistors in series and parallel.

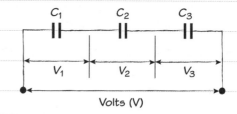

Capacitors in series where charge (Q) is a constant.

Capacitors in parallel

When capacitors are connected in **parallel** there is the **same voltage** across each capacitor.

The total capacitance (C_T) for a number (n) of capacitors in parallel is:

$C_T = C_1 + C_2 \ldots + C_n$

Capacitors in parallel where voltage (V) is constant.

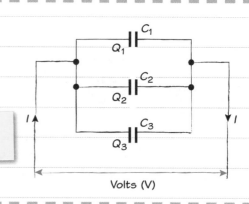

Worked example

A simple circuit contains three capacitors connected in **series**, where $C_1 = 2\,\mu F$, $C_2 = 4\,\mu F$ and $C_3 = 5\,\mu F$. The circuit supply voltage $V = 110\,V$ DC.

Calculate the charge stored in each capacitor.

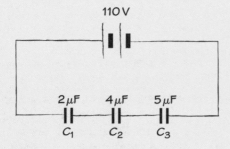

Sample response extract

For capacitors in series

$\frac{1}{C_T} = \frac{1}{C_1} + \frac{1}{C_2} + \frac{1}{C_3} = \frac{1}{2} + \frac{1}{4} + \frac{1}{5}$

$C_T = \frac{1}{0.95} = 1.052\ldots\ \mu F$

Charged stored:

$Q = CV$

$Q = 1.052\ldots \times 10^{-6} \times 110 = 1.16 \times 10^{-4}\,C$

Each capacitor holds the same charge.

Follow these steps:
1 Annotate the circuit diagram.
2 Find the combined total capacitance of the series capacitors.
3 Use $Q = CV$ to find the charge required.
4 Remember that in a series circuit the charge stored in each capacitor is the same.

Now try this

A simple circuit contains three capacitors connected in **parallel**, where $C_1 = 2\,\mu F$ and $C_2 = 4\,\mu F$, $C_3 = 5\,\mu F$. The circuit supply current $V = 110\,V$ DC. Calculate the charge stored in each capacitor.

Capacitors in series and parallel combinations

It is common for electrical and electronic circuits to contain capacitors connected together in series and parallel combinations. It's important that you understand the characteristics of these networks.

Series and parallel combinations

First calculate the equivalent resistance of the parallel element, using:

$$C_P = C_1 + C_2$$

Then use:

$$\frac{1}{C_T} = \frac{1}{C_P} + \frac{1}{C_3}$$

to find the total capacitance of the combination.

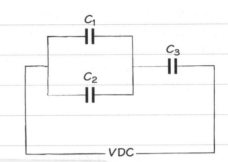

Capacitors connected in a series and parallel combination.

Worked example

A circuit, as shown, contains three capacitors in a series/parallel combination, where $C_1 = 2\,\mu F$, $C_2 = 4\,\mu F$ and $C_3 = 5\,\mu F$. The circuit supply voltage $V = 110\,V$ DC.

Calculate the total stored charge in the circuit.

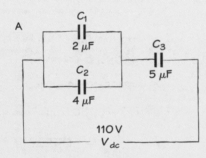

Sample response extract

Simplifying parallel combination:

$$C_A = C_1 + C_2 = 2 + 4 = 6\,\mu F$$

Combining series elements of simplified circuit:

$$\frac{1}{C_T} = \frac{1}{C_A} + \frac{1}{C_3} = \frac{1}{6} + \frac{1}{5} = \frac{11}{30}$$

$$C_T = 2.72\,\mu F$$

Charge stored:

$$Q = CV$$

$$Q = 2.72 \times 10^{-6} \times 110 = 2.99 \times 10^{-4}\,C$$

Follow these steps:

1. Identify the parts of the circuit that can be simplified and annotate the circuit diagram.
2. Find the combined total capacitance of C_1 and C_2 in parallel. Call this C_A.
3. Find the combined total capacitance of C_A and C_3 in series.
4. Use $Q = CV$ to find the charge.

Continue with the Worked example and find the charge stored in capacitor C_1.

Capacitors in series carry the same charge. Find the voltage across the parallel combination using $V = Q/C$. Find the charge in C_1 using $Q = CV$.

Magnetism and magnetic fields

Magnetism underpins a huge range of important areas of engineering, such as electrical generators, motors, transformers and a host of other electro-mechanical devices.

Magnetic fields

Magnetic flux (Φ) is used to measure the total magnetic field produced by a source of magnetism. Units webers (Wb).

Magnetic flux density

Magnetic flux density (the intensity of a magnetic field), in Wb/m² or tesla (T)

$$B = \frac{\Phi}{A}$$

Magnetic flux, in Wb

Area, in m²

Units Wb/m² or Tesla (T).

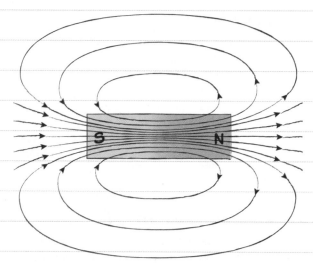

Lines of magnetic flux around a bar magnet.

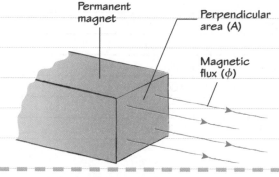

Permanent magnet

Perpendicular area (A)

Magnetic flux (ϕ)

Lines of magnetic flux from the end of a bar magnet.

Ferromagnetic materials

- iron and steel
- can be magnetised by a permanent magnet or the magnetic field of a solenoid.

Solenoids

Magneto motive force (F_m)

Magneto motive force (A)

$$F_m = NI$$

Number of turns

Electrical current through a solenoid (A)

Units A.

Magnetic field strength (H)

Magnetic field strength (A/m)

$$H = \frac{NI}{L}$$

Number of turns

Electrical current through a solenoid (A)

Length (m)

Units A/m¹.

Length of solenoid (L)

Magnetic field strength (H)

Number of wire turns in solenoid (N)

Current flowing in solenoid (I)

Now try this

A solenoid is 25 mm in length and has 250 turns. When connected to a power supply, a current of 1.2 A flows through the solenoid.

Calculate the magnetic field strength generated inside the solenoid.

Permeability

Permeability is a measure of the degree of magnetisation a material undergoes when it is exposed to a magnetic field.

Permeability (μ)

The ratio of the magnetic flux density (**B**) generated inside a material and the external magnetic field strength (**H**) that causes the magnetising effect.

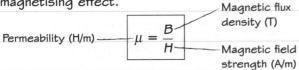

Permeability (H/m) —— $\mu = \dfrac{B}{H}$

— Magnetic flux density (T)

— Magnetic field strength (A/m)

Units H/m.

Permeability of free space (μ_o)

The theoretical ratio of magnetic flux density (**B**) and magnetising field strength (**H**) in a vacuum. This is a fixed constant and used as a benchmark to which other materials are compared.

$\mu_O = 4\pi \times 10^{-7}$ H/m

The unit is the henry per metre.

Relative permeability (μ_r)

This compares the permeability (μ) observed in a given material to the permeability of free space (μ_O).

$\mu_r = \dfrac{\mu}{\mu_O}$

This relationship can also be expressed as:

$\dfrac{B}{H} = \mu_O \mu_r$

Medium	Relative permeability (μ_r)
Vacuum	1
Air	1.000 000 37
Wood	1.000 000 43
Aluminium	1.000 022
Carbon steel	100
Electrical steel	4000
Iron	5000
Permalloy	8000

Values of the relative permeability (μ_r) of some common materials.

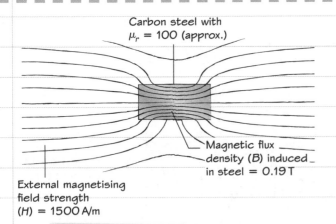

Carbon steel with $\mu_r = 100$ (approx.)

Magnetic flux density (**B**) induced in steel = 0.19 T

External magnetising field strength (**H**) = 1500 A/m

In carbon steel the magnetising field $H = 1500$ A/m causes a magnetic flux density $B = 0.19$ T.

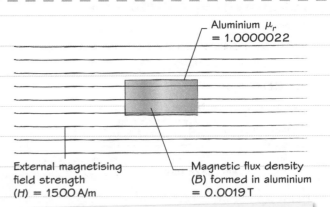

Aluminium $\mu_r = 1.0000022$

External magnetising field strength (**H**) = 1500 A/m

Magnetic flux density (**B**) formed in aluminium = 0.0019 T

The same magnetising field in aluminium causes a much lower magnetic flux density B = 0.0019 T, which is little more than exists in the surrounding air.

Now try this

A bar of electrical steel with $\mu_r = 4000$ is placed inside a solenoid and exposed to a magnetic field strength which is $H = 2200$ A/m.

The permeability of free space, μ_{0r} is $4\pi \times 10^{-7}$ H/m.

Calculate the magnetic flux density (B) formed in the electrical steel.

B/H curves, loops and hysteresis

It is important that you are aware of the relationship between the magnetic field strength (H) and magnetic flux density (B) in ferromagnetic materials where permeability is not a constant.

B/H curves in ferromagnetic materials

A B/H curve is simply a graph of the magnetic flux density (B) and applied magnetic field strength (H) for a given material.

These are not straight lines as you might expect. They are curves because permeability (μ) decreases as the applied magnetic field (H) increases.

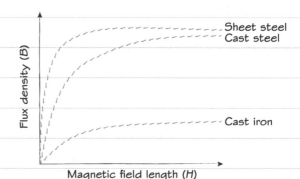

Typical B/H curves for a range of ferromagnetic materials.

At low values of H where μ is high, a small increase in H leads to a large increase in B.
At high values of H, where the material approaches magnetic saturation, μ tends to 0 and an increase in H leads to no further increase in B.

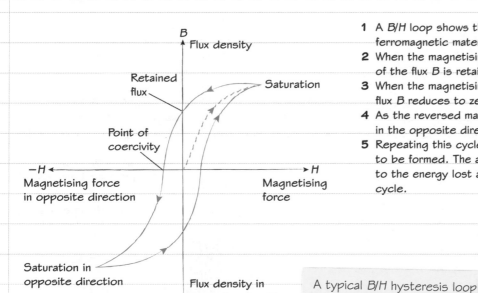

A typical B/H hysteresis loop for a ferromagnetic material.

1 A B/H loop shows the full magnetisation of a ferromagnetic material along the dotted line.
2 When the magnetising field H is reduced to zero, some of the flux B is retained.
3 When the magnetising field H is reversed, the retained flux B reduces to zero at the point of coercivity.
4 As the reversed magnetic field H increases, saturation in the opposite direction is reached.
5 Repeating this cycle causes a closed hysteresis loop to be formed. The area inside this loop is proportional to the energy lost as heat during the magnetisation cycle.

Now try this

In many applications, ferromagnetic materials undergo the magnetisation cycle many times a second.

Explain the importance of specifying a ferromagnetic material with a narrow hysteresis loop when designing a transformer core.

Reluctance and magnetic screening

Some materials oppose the flow of magnetic flux. Reluctance (S) is used to describe this magnetic resistance. Units are 1/H.

Analogy of reluctance and resistance

You can think of reluctance as being similar to electrical resistance:

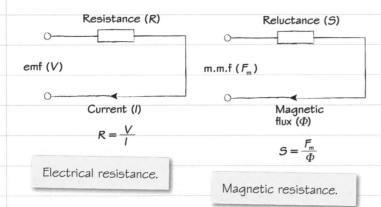

$$R = \frac{V}{I}$$

Electrical resistance.

$$S = \frac{F_m}{\Phi}$$

Magnetic resistance.

Reluctance (S) can be expressed as the ratio of magneto motive force (F_m) to magnetic flux (Φ):

$$S = \frac{F_m}{\Phi}$$

Reluctance (S)

You can also express reluctance in terms of the characteristics of the material used in a magnetic circuit.

$$S = \frac{l}{\mu_0 \mu_r A}$$

μ_0 = permeability of free space

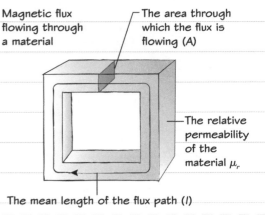

Magnetic flux flowing through a material

The area through which the flux is flowing (A)

The relative permeability of the material μ_r

The mean length of the flux path (l)

Magnetic screening

Often it is necessary to screen sensitive electronic components or devices from magnetic fields to prevent any unwanted effects. Effective screening can be achieved using ferromagnetic materials with low reluctance. These provide a pathway for lines of magnetic flux around the components being protected.

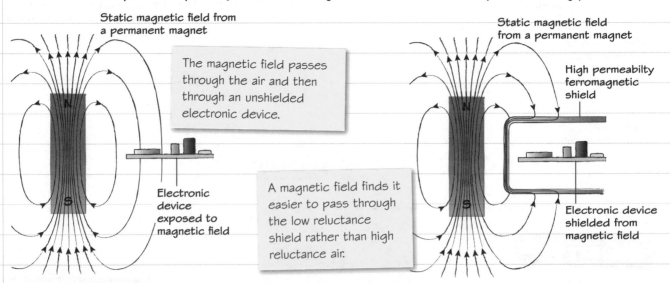

Static magnetic field from a permanent magnet

The magnetic field passes through the air and then through an unshielded electronic device.

Electronic device exposed to magnetic field

Static magnetic field from a permanent magnet

High permeabilty ferromagnetic shield

A magnetic field finds it easier to pass through the low reluctance shield rather than high reluctance air.

Electronic device shielded from magnetic field

Now try this

A steel circular core with relative permeability 190 has a mean length of 0.2 m and a cross-sectional area of 0.01 m². The permeability of free space is $4\pi \times 10^{-7}$ H/m.

Calculate the magnetic reluctance of the circuit.

Electromagnetic induction

Electromagnetic induction is the key process in the generation of electricity.

Faraday's law of electromagnetic induction

The magnitude of the emf generated is given by Faraday's law:

'When a magnetic flux through a coil is made to vary, an emf is induced. The magnitude of this emf is proportional to the rate of change of flux.'

Lenz's law of electromagnetic induction

The direction of the induced current is given by Lenz's law:

'An induced current always acts in such a direction so as to oppose the change in flux producing the current.'

The emf induced in a coil

An emf is generated in a stationary coil by the movement of a permanent magnet, which provides the changing magnetic flux.

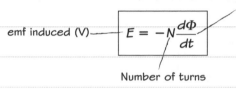

emf induced (V) —— $E = -N\dfrac{d\Phi}{dt}$ —— Rate of change of magnetic flux (the −ve sign is a consequence of Lenz's law)

Number of turns

A simple experiment can be carried out to demonstrate electromagnetic induction in a coil.

 S

Magnet made to move through centre of coil

N

Stationery coil with number of turns, *N*

Voltmeter indicating size of induced emf

V

The emf induced in a moving conductor

An emf is generated in a conductor moving through a stationary magnetic field at right angles to the lines of flux.

emf induced (V) —— $E = BLv$ —— Velocity (m/s)

Flux density (T)

Length (m)

emf induced in a wire moving in a magnetic field.

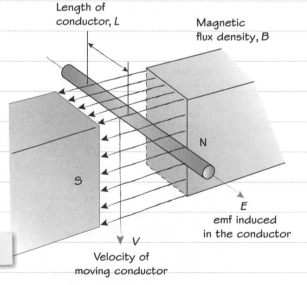

Length of conductor, *L*

Magnetic flux density, *B*

N

S

E emf induced in the conductor

V Velocity of moving conductor

A conductor moving at right angles to a magnetic field with a flux density of 0.04 T at a velocity of 0.5 m/s generates an emf of 2.4×10^{-3} V.

Calculate the length of the conductor moving within the field.

DC motors

Motors convert electrical energy to mechanical rotation able to do work.

The basic operation of a DC motor

- A permanent magnet provides a static magnetic field in which a coil is free to rotate.
- When current flows in the coil, the magnetic field generated opposes the static field from the magnets and the coil rotates until the fields are aligned.
- As this happens, the commutator reverses the current in the coil.
- This, in turn, reverses its magnetic field, causing the rotation to continue.

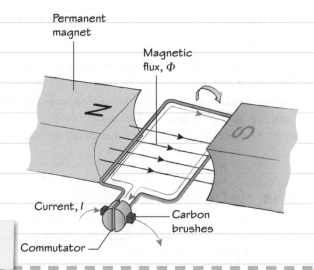

Permanent magnet

Magnetic flux, Φ

Current, I

Carbon brushes

Commutator

The DC motor. Current is fed through carbon brushes to a split ring, commutator, which reverses the current every half-turn.

Industrial DC motors

In practice, motors are more refined. Some features of industrial motors include:

1. The rotating section, including the coils (or windings), commutator and drive shaft, is called an **armature** and is supported in bearings to allow free rotation.

2. The commutator is split into several segments, feeding current to several coils or windings in turn to provide smooth operation.

3. The stationary part of the motor, or **stator**, uses electromagnets to provide the static magnetic field, because permanent magnets have a limit to the size of magnetic field they can produce.

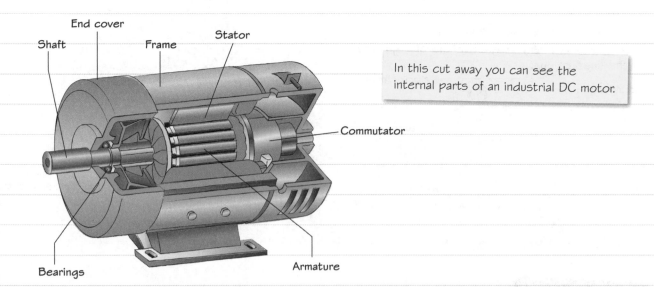

End cover

Shaft

Frame

Stator

Commutator

Bearings

Armature

In this cut away you can see the internal parts of an industrial DC motor.

Now try this

Explain the function of the commutator and the armature in a DC electric motor.

Electric generators

Electric generators are used to convert mechanical power, usually provided by a rotating turbine or engine, to electricity.

Operation of an electric generator

- Generators rely on the principles of electromagnetic induction.
- A coil rotates inside a static magnetic field, here provided by a permanent magnet.
- An emf is induced in the coil.
- Each end of the coil is connected to two separate slip rings, which maintain contact with carbon brushes.
- The induced emf causes an electric current to flow between the brushes as the coil rotates.

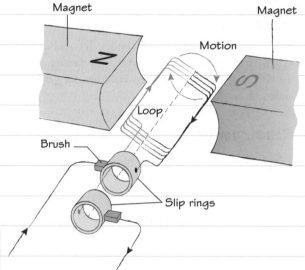

This sketch illustrates the basic construction of an electric generator.

Factors effecting induced emf

The magnitude of the emf generated is proportional to:

- the number of turns in the coil
- the speed of rotation
- the strength of the magnetic field in which the coil rotates
- the angle of the coil relative to the magnetic field in which it is rotating.

Max emf induced when coil is cutting through perpendicular lines of magnetic flux.

Min emf (V = 0) when coil is moving parallel to lines of magnetic flux.

Sinusoidal output

As the induced emf at any time depends on the angle at which the coil is cutting through the lines of magnetic flux, the output from this type of generator is sinusoidal.

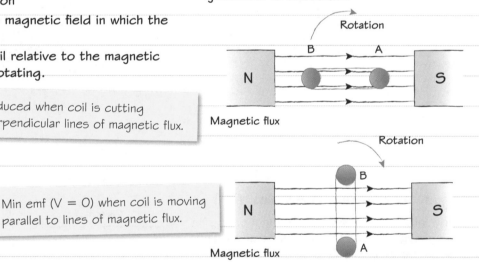

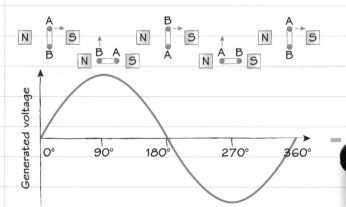

Over a complete rotation of the coil a single cycle of a sinusoidal waveform is generated. This waveform is better known as an alternating current (AC) and for this reason generators of this type are often called **alternators**.

Now try this

State three factors that increase the emf generated by an alternator.

Inductors and self-inductance

Self-inductance occurs when the varying magnetic field generated by a change in the current flowing in a circuit induces an emf in the same circuit.

Inductance (*L*)

Inductance is measured in henry (H), whereby 1H is the inductance present in a circuit where a changing current of 1 A/s induces an emf of 1V. Units H.

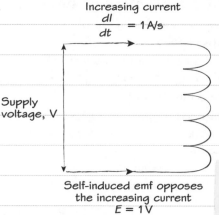

Increasing current
$$\frac{dI}{dt} = 1 \text{ A/s}$$

Supply voltage, V

Circuit inductance
$L = 1\,H$

Self-induced emf opposes the increasing current
$E = 1\,V$

A circuit with self-inductance of 1H.

Electromotive force or emf (*E*)

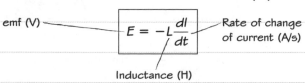

emf (V) — $E = -L\dfrac{dI}{dt}$ — Rate of change of current (A/s)

Inductance (H)

Units V.

Self-inductance in a coil (*L*)

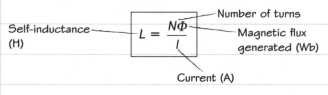

Self-inductance (H) — $L = \dfrac{N\Phi}{I}$ — Number of turns

Magnetic flux generated (Wb)

Current (A)

Units H.

A coil with self inductance L.

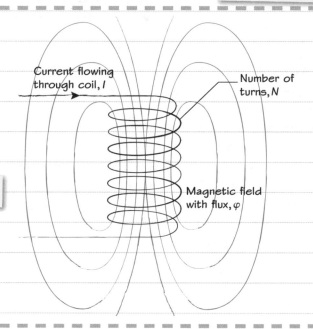

Current flowing through coil, I

Number of turns, N

Magnetic field with flux, φ

Energy stored in an inductor (*W*)

When a flow of current is established in a conductor a magnetic field is also generated. The work done in creating this field is stored within it until the current stops flowing. As the current flow ceases the collapse of the magnetic field returns the stored energy as an induced emf.

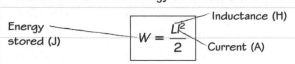

Energy stored (J) — $W = \dfrac{LI^2}{2}$ — Inductance (H)

Current (A)

Units J.

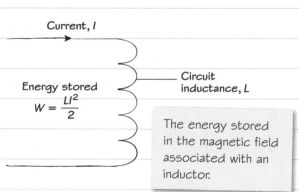

Current, I

Energy stored
$$W = \frac{LI^2}{2}$$

Circuit inductance, L

The energy stored in the magnetic field associated with an inductor.

Now try this

An inductor with inductance 2.2 H passes a current of 0.3 A.
Calculate the energy stored in the inductor.

Transformers and mutual inductance

Transformers use the principles of mutual inductance to change the voltage and current characteristics of an AC supply.

Mutual inductance (M)

Mutual inductance occurs when the varying magnetic field generated by a change in the current flowing in one circuit induces an emf in an adjacent circuit.

emf induced (V) —
$$E = -M\frac{dI}{dt}$$
— Mutual inductance (H)
— Rate of change of current (A/s)

A circuit with mutual inductance of 1H.

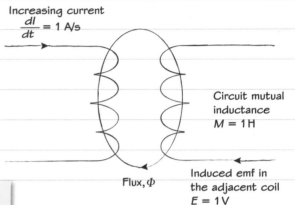

Increasing current
$$\frac{dI}{dt} = 1\,A/s$$

Circuit mutual inductance
$M = 1\,H$

Flux, Φ

Induced emf in the adjacent coil
$E = 1\,V$

Transformers

A transformer

- has two separate coils or windings
- laminated iron core reduces losses through eddy currents
- iron core minimises hysteresis losses.

Typical transformers.

Important transformer calculations

For a primary coil with turns N_1 and voltage V_1 and a secondary coil with turns N_2 and voltage V_2, the transformer voltage ratio is given by:

$$\frac{V_1}{V_2} = \frac{N_1}{N_2}$$

The transformer current ratio is given by:

$$\frac{V_1}{V_2} = \frac{I_2}{I_1}$$

A step-up transformer increases the supply voltage.

A step-down transformer reduces the supply voltage.

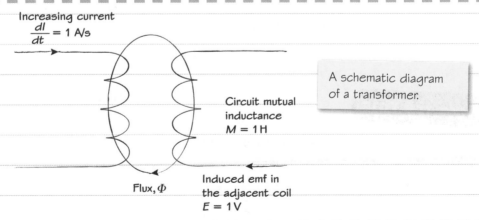

Increasing current
$$\frac{dI}{dt} = 1\,A/s$$

Circuit mutual inductance
$M = 1\,H$

A schematic diagram of a transformer.

Flux, Φ

Induced emf in the adjacent coil
$E = 1\,V$

Now try this

A transformer with a primary supply voltage of 230V AC has a primary winding with 1200 turns and a secondary winding with 500 turns.

Calculate the voltage output from the secondary winding.

AC waveforms

An alternating current can be analysed by looking at how voltage and current vary over time. The resulting graphs are known as waveforms.

Sinusoidal waveforms

- the output from electrical alternators (generators)
- all mains electricity.

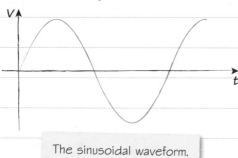

The sinusoidal waveform.

Square

Triangular

Sawtooth

Non-sinusoidal waveforms.

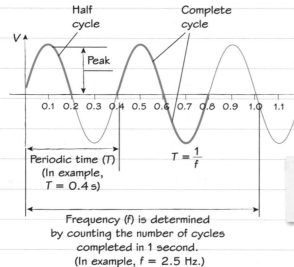

Half cycle

Complete cycle

Peak

Periodic time (T)
(In example, T = 0.4 s)

$$T = \frac{1}{f}$$

Frequency (f) is determined by counting the number of cycles completed in 1 second.
(In example, f = 2.5 Hz.)

The sinusoidal waveform is periodic, being made up of a series of repeating cycles.

Now try this

A sinusoidal waveform has a periodic time of 0.016 s.
Calculate the frequency of the waveform.

Single phase AC parameters

Different parameters are used to define the characteristics of a single phase AC supply.

AC parameters

- **Peak voltage (V_{peak})**

The maximum value of voltage or current reached in a positive or negative half cycle.

(For the example shown below V_{peak} = 325 V.)

- **Peak-to-peak voltage**

The difference between the positive peak and the negative peak voltage in a full cycle.

(For the example shown below peak to peak voltage = 650 V.)

- **Root mean square (rms) voltage (V_{rms})**

The equivalent DC voltage and how AC voltages are usually quoted. For a sine wave:

$$V_{rms} = \frac{1}{\sqrt{2}} V_{peak}$$

(For the example shown below V_{rms} = 230 V.)

- **Average voltage (V_{ave})**

The average of all the instantaneous measurements in one half cycle. For a sine wave:

$$V_{ave} = \frac{2}{\pi} V_{peak}$$

(For the example shown below V_{ave} = 207 V.)

- **Form factor**

$$\text{Form factor} = \frac{V_{rms}}{V_{ave}}$$

For a sinusoidal waveform the form factor is a constant 1.11.

Note: These relationships are also true when considering current instead of voltage.

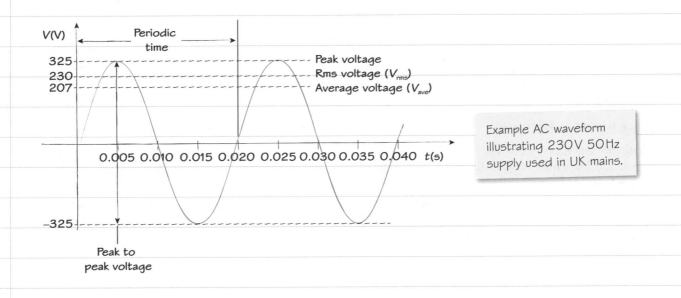

Example AC waveform illustrating 230 V 50 Hz supply used in UK mains.

Now try this

Building sites use 110V AC to power portable equipment, as this is less hazardous than the standard 230V AC mains supply.

Calculate the peak voltage for a 110V AC supply.

Analysing AC voltages using phasors

You can represent a sinusoidal AC voltage as a sine wave on a graph. Another useful tool is to think of this wave in terms of a rotating vector or phasor.

As a phasor rotates, its vertical height at any instant corresponds to the voltage at that point on the corresponding sine wave.

$$V = V_{peak}\sin(\omega t + \phi)$$

Voltage (in V) at time t (s)

Phase angle (rad)

$$\omega = 2\pi f$$

Angular velocity (rad/s)   Frequency (Hz)

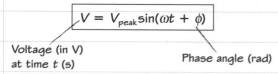

Direction of vector rotation

$\theta = \omega t$

$\theta = \omega t$

V_{peak}

Rotating phasor with angular velocity ω rad/s

Corresponding sinusoidal waveform, where $V = V_{peak}\sin(\omega t + \phi)$

Using phasors to represent a sinusoidal waveform.

At $t = 0$
$V = 0$ (Phase angle $\phi = 0$)

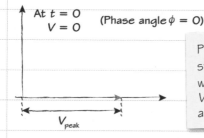

V_{peak}

Phasor representing standard sinusoidal waveform, where $V = V_{peak}\sin(\phi t + \phi)$ and $\phi = 0$.

At $t = 0$
$V \neq 0$ (Phase angle $\phi \neq 0$)

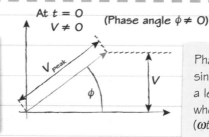

V_{peak}

ϕ

V

Phasor representing a sinusoidal waveform with a leading phase angle, where $V = V_{peak}\sin(\omega t + \phi)$ and $\phi \neq 0$.

Graphical addition

Sinusoidal waveforms can be added together graphically. However, this is laborious and time consuming.

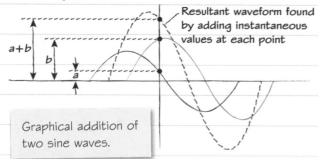

$a+b$

b

a

Resultant waveform found by adding instantaneous values at each point

Graphical addition of two sine waves.

Phasor addition

Sinusoidal waveforms of the same frequency can be added together using the vector addition of phasors.

For example, adding $v_1 = 100\sin(100\omega t)$ and $V_2 = 200\sin(100\omega t + \frac{\pi}{6})$ can be represented as:

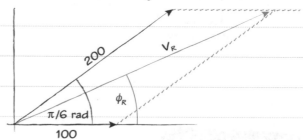

200

V_R

ϕ_R

$\pi/6$ rad

100

The resultant peak voltage V_R and phase angle ϕ_R can be determined using basic trigonometric methods.

Links Trigonometric methods for carrying out vector addition can be found on page 13.

Links Trigonometric methods for carrying out vector addition can be found on page 13.

Links For an example of how to solve phasor addition problems graphically, see page 85.

Now try this

Two sinusoidal voltages of the same frequency are represented by the equations $V_1 = 100\sin(100\omega t)$ and $V_2 = 200\sin(100\omega t + \frac{\pi}{6})$.

If these are combined, determine the equation for the resultant waveform.

Reactance and impedance

In AC circuits the equivalent to resistance is impedance (Z), which is a combination of conventional resistance (R), capacitive reactance (X_C) and inductive reactance (X_L).

Capacitive reactance (X_C)

Capacitive reactance is the opposition to the flow of alternating current exhibited by a capacitor.

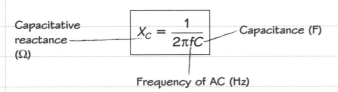

Capacitive reactance (Ω) —— $X_C = \dfrac{1}{2\pi fC}$ —— Capacitance (F)

Frequency of AC (Hz)

Inductive reactance (X_L)

Inductive reactance is the opposition to the flow of alternating current exhibited by an inductor.

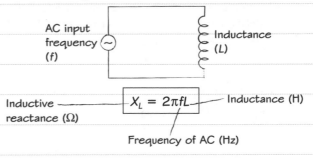

Inductive reactance (Ω) —— $X_L = 2\pi fL$ —— Inductance (H)

Frequency of AC (Hz)

Resistor/capacitor series circuit

In circuits containing a capacitor and a resistor in series then the total impedance (Z) is made of the resistance (R) and capacitive reactance (X_C).

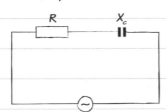

Resistor/inductor series circuit

In circuits containing an inductor and a resistor in series then the total impedance (Z) is made up of the resistance (R) and the inductive reactance (X_L).

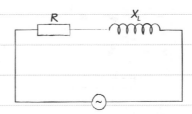

Total impedance of a resistor/capacitor series circuit

In AC circuits resistance (R) and capacitive reactance (X_C) are vector quantities.

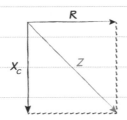

Total impedance (Z) is given by:

$$Z = \sqrt{X_C^2 + R^2}$$

Total impedance of a resistor/inductor series circuit

In AC circuits resistance (R) and inductive reactance (X_L) are vector quantities.

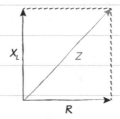

Total impedance (Z) is given by:

$$Z = \sqrt{X_L^2 + R^2}$$

Now try this

An engineer is testing an inductor with an inductance of 1.2 H and resistance 2.5 Ω. It is connected to a 230 V, 50 Hz AC supply.

(a) Calculate the impedance of the circuit.

(b) Calculate the current drawn from the supply.

Rectification

An important application of diodes is in the rectification of AC to DC. Rectifier circuits are commonly used to convert the low voltage AC output of a mains transformer to a DC voltage suitable for running electronic devices.

Simple half wave rectifier

A simple circuit connected to the output of a transformer uses a single diode to provide half wave rectification by eliminating the -ve half cycle.

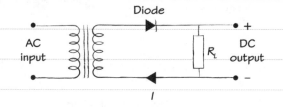

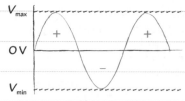

AC input.

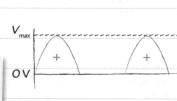

DC output with –ve half cycle eliminated.

Full wave bridge rectifier

A circuit connected to the output of a transformer uses a 4-diode bridge to provide full wave rectification making both half cycles +ve.

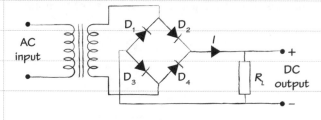

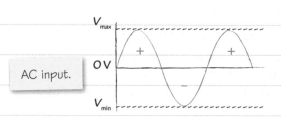

AC input.

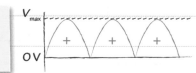

DC output with –ve half cycle inverted to become +ve.

Smoothed full wave bridge rectifier

The addition of a smoothing capacitor to the output of the rectifier reduces voltage ripple and provides a smoother DC output.

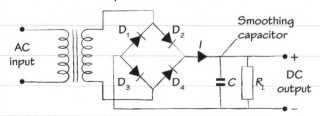

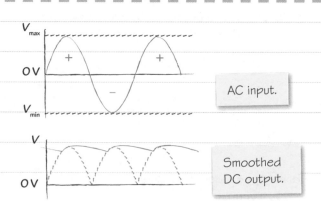

AC input.

Smoothed DC output.

Now try this

Electricity distribution systems use high voltage AC (alternating current) to carry electricity over long distances between power stations and homes and businesses. However, many electronic devices require a much lower voltage DC (direct current) electrical supply in order to function.

Use notes and diagrams to explain how 240V AC mains electricity could be converted to a 5V DC supply suitable for use in a phone charger.

Your Unit 1 exam

Your Unit 1 exam will be set by Pearson and could cover any of the essential content in the unit. You can revise the unit content in this Revision Guide. This skills section is designed to revise skills that might be needed in your exam.

Exam checklist

Before your exam, make sure you:

☑ Have double-checked the time and date of your exam.

☑ Have a black pen you like and at least one spare.

☑ Have checked what you need to take into the exam; e.g. a ruler, protractor, pencil, scientific calculator that must not be programmable and that meets the requirements stated.

☑ Are prepared to show all your working using the appropriate units in your answers, to the appropriate degree of accuracy.

☑ Get a good night's sleep.

Check the Pearson website

The questions provided in this section are designed to demonstrate the skills that might be needed in your exam. The details of the actual exam may change so always make sure you are up to date. Check the Pearson website for the most up-to-date **Sample Assessment Material** to understand the structure of your paper and how much time you are allowed.

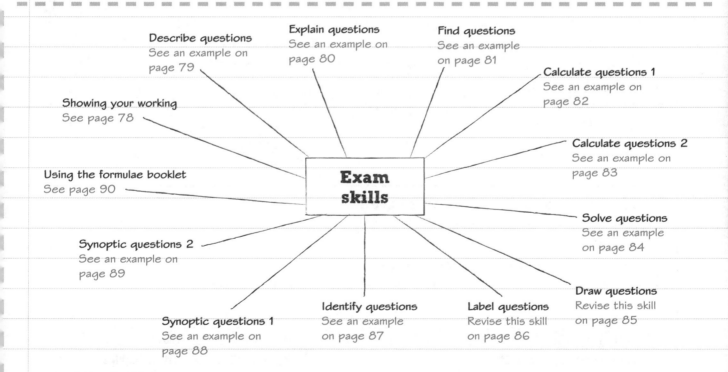

Describe questions
See an example on page 79

Explain questions
See an example on page 80

Find questions
See an example on page 81

Calculate questions 1
See an example on page 82

Showing your working
See page 78

Calculate questions 2
See an example on page 83

Using the formulae booklet
See page 90

Exam skills

Synoptic questions 2
See an example on page 89

Solve questions
See an example on page 84

Synoptic questions 1
See an example on page 88

Identify questions
See an example on page 87

Label questions
Revise this skill on page 86

Draw questions
Revise this skill on page 85

Now try this

Visit the Pearson website and find the page containing the course materials for BTEC National Engineering. Look at the latest Unit 1 Sample Assessment Material (SAM) to get an indication of:

- the number of papers you have to take
- whether a paper is in parts
- how much time is allowed and how many marks are allocated
- what types of questions appear on the paper.

Your tutor or instructor may already have provided you with a copy of the Sample Assessment Material. You can use these as a 'mock' exam to practise before taking your actual exam.

Showing your working

Here are some examples of the skills involved when showing your workings and methods in your mathematical calculations.

Worked example

The supports for the loaded beams of a bridge must be able to support the load safely.

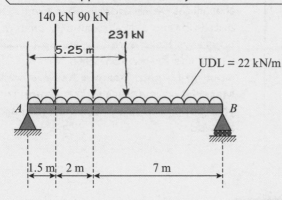

Calculate the reaction force at point *A*.   **2 marks**

> If answering, for example, a simply supported beam question that includes a UDL (uniformly distributed load), add the equivalent point load to the sketch provided.

> You should set out your working clearly and, where necessary, explain in words what you are doing at each stage.

Sample response extract

Find equivalent point force for UDL.

Magnitude of force:

$(1.5 + 2 + 7) \times 22 = 231$ kN

Position of force:

$\dfrac{(1.5 + 2 + 7)}{2} = 5.25$ m from point A

Find support reaction at point A.

Take moments about B:

$R_A \times 10.5 = (140 \times 9) + (90 \times 7) + (231 \times 5.25)$

$R_A \times 10.5 = 1260 + 630 + 1112.75$

$R_A = \dfrac{3002.75}{10.5}$ ← Numerical error

$R_A \times 285.98$ kN

> The response to this question clearly shows the method followed. However, in this case, the final solution is incorrect because of a numerical error made in one of the calculations.

> Despite the fact that the final answer is inaccurate, the correct method is clearly shown. So this would be reflected in the marking, as shown below.

Final answers, methods and marks

In your exam, marks may be given for:

☑ your final answer

☑ the method used to reach the final answer.

For example: 1 mark for method   1 mark for answer

Working	Answer	Notes
Taking moments about B: $R_A \times 10.5 = (140 \times 9) + (90.7 \times 7) + (281 \times 5.25)$ $R_A \times 10.5 = 1260 + 630 + 1212.75$ $R_A = \dfrac{3102.75}{10.5}$ $R_A = 295.5$ kN	$R_A = 295.5$ kN	M1 for taking moments about B. A1 for value of R_A.

Now try this

Continue with the Worked example to calculate the reaction force at point *B*.

🔗 **Links**   You could revise page 18 to help you.

'State' and 'describe' questions

Here are some examples of skills involved when answering '**state**' and '**describe**' questions.

Worked example

A body in static equilibrium will remain at rest or in motion with constant velocity.

State all three conditions that must be met for a system of coplanar non-concurrent forces to be in static equilibrium.

1 mark

If answering a 'state' question, recall the specific facts required and state them clearly in your answer.

You need to correctly state all three of the conditions that must be met.

Sample response extract

The sum of all the vertical components of the forces acting in the system is zero.

The sum of all the horizontal components acting in the system is zero.

The sum of all the moments acting about any point in the system is zero.

Links Look at pages 16 to revise the content covered in this question.

Worked example

A maintenance engineer strips down a DC electric motor in order to find and rectify a fault that is causing it to malfunction. They have traced the fault to worn brushes, which will need replacing.

Describe the function of the brushes in a DC electric motor.

2 marks

The first part of the question establishes the **engineering context** of the question so that you will know the subject area being considered.

When answering a 'describe' question, you need to give a clear description of a particular component, feature or process. Here, the answer is structured in sentences and provides a good description of the function of the brushes.

Sample response extract

A pair of stationary carbon brushes allow electrical current to be supplied to the coils in the rotating armature of the motor. They are spring-loaded and stay in contact with the commutator, which feeds current to each coil in turn as the motor spins.

Brushes are usually made of carbon and wear down gradually over the lifetime of the motor. Worn brushes are a common fault.

Whenever possible you should use appropriate technical language in your answers. This example correctly uses the terms **armature** and **commutator** in connection with the function of brushes.

The second paragraph of this answer gives an indication of a good understanding of the practical aspects of using electric motors. However, you are unlikely to have time to add in extra detail in this way, so concentrate your efforts and write concise answers.

Now try this

A vacuum flask helps to keep hot things hot and cold things cold by limiting heat transfer between the inside and outside of the flask.

State the three principal mechanisms of heat transfer.

Links You can revise heat transfer mechanisms on page 33.

Electrical generators or alternators are widely used to generate electricity in a wide range of applications.

Describe the function of the slip rings in a simple electric generator.

'Explain' questions

Here are some examples of skills involved when answering **explain** questions.

Worked example

Electric motors are used widely to convert electrical energy into mechnical energy to run everying from extractor fans to trains. They are clean, reliable and able to operate at efficiencies of up to 90%.

Explain one form of mechanical loss that limits the efficiency of an electric motor. **2 marks**

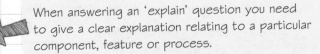

When answering an 'explain' question you need to give a clear explanation relating to a particular component, feature or process.

Sample response extract

Mechanical losses in electric motors are mainly due to friction. This occurs wherever surfaces move over one another, such as in the bearings that allow the armature of the motor to rotate. Friction losses take the form of waste heat.

The answer is structured in two useful parts.
1 The first sentence identifies a source of mechanical loss.
2 The rest of the answer expands on this to explain its cause, characteristics and effects.

Always write as neatly as possible and make sure your answers **make sense** when you read them back to yourself. They should be easy to read and understand.

Remember to use appropriate **technical language** in your answers wherever it is appropriate.

Answering the question

'**Explain**' questions may require answers that are fairly short, as in this example, or a bit longer.

For example, you may be asked to explain two components, features or processes, instead of just one. If so, you can explain each one using the same approach shown here.

Now try this

Heat energy can be transferred from one place to another in a range of different ways.

Explain the primary form of heat transfer used in a household hairdryer.

'Find' questions

Here are some examples of skills involved when answering **find** questions.

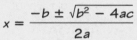 **Worked example**

When answering a '**find**' question, determine the solution to a problem given in the information provided in the question. This might involve applying the particular technique or mathematical method mentioned in the question.

The radius, r, of an enclosed cylindrical tank with height of 0.5 m and total surface area of 5 m^2 is given by the equation:

$5 = 2\pi r^2 + \pi r$

A question may have a short scenario giving the **engineering context** for the question.

Use the quadratic formula to find the radius of the tank.

2 marks

In this example, the equation has been recognised as a **quadratic equation**. The first step in finding the solution is to rearrange the equation into its standard form, which can be found in the formula booklet.

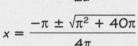 **Sample response extract**

$0 = 2\pi r^2 + \pi r - 5$

$y = ax^2 + bx + c$

$a = 2\pi \qquad x = r$

$b = \pi$

$c = -5$

Writing down the values of coefficients a, b and c helps to **make the working clear**. This is helpful when giving an answer.

$x = \dfrac{-b \pm \sqrt{b^2 - 4ac}}{2a}$

In this example, a reminder of the quadratic formula would be found in the **formula booklet** and copied into the solution, then substituting in the values for a, b and c.

$x = \dfrac{-\pi \pm \sqrt{\pi^2 + 40\pi}}{4\pi}$

$x = \dfrac{-\pi \pm 11.639}{4\pi}$

$x = 0.676$ or -1.176

The quadratic formula will generate two possible solutions. You would need to decide whether one or both of these are relevant in the context of the question. In this case, it is not possible to have a negative radius and so this possible solution would be ignored.

Negative answers are not possible in this scenario.

Required radius = 0.676 m

You need to state your **final answer** clearly to an appropriate level of accuracy. This will usually be to 2 d.p. or 3 s.f.

If time allows, always **check your solution** by substituting it back into the equation given in the question.

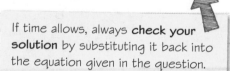 **Now try this**

The radius, r, of an enclosed cylindrical tank with height of 1 m and total surface area of 16 m^2 is given by the equation:

$16 = 2\pi r^2 + 2\pi r$

Use the quadratic formula to find the radius of the tank.

'Calculate' questions 1

Here are some examples of skills involved if answering 'calculate' questions where you need to find the number or amount of something based on information given in the question.

Worked example

The diagram shows a radio mast supported by a triangular system of cables. An engineer has made a ground level survey of the mast supports and measured the angles and distances shown in the diagram.

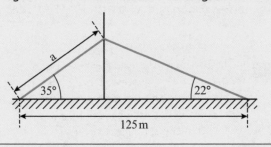

Calculate the length of cable *a*.

2 marks

> The question has a short scenario giving the **engineering context** for the calculation, and you need to focus on the application of a mathematical method or formula.

> You may find a small **diagram** is included to illustrate what is required in the question.

Sample response extract

In this case:

B = 35°

A = 22°

C = 125 m

Angle C = 180 − 35 − 22

C = 123°

Using the sine rule:

$$\frac{a}{\sin A} = \frac{c}{\sin C}$$

Rearrange:

$$a = \frac{\sin A \times c}{\text{Sin} C}$$

Substitute known values:

$$a = \frac{\sin 22 \times 125}{\text{Sin} 123}$$

a = 55.83 (to 2 d.p)

The length of cable a is 55.83 m.

> You might find it useful to add your own labels or notes to a small **sketch** or the diagram itself as you work. In this example, labels have been added to help identify the sides and angles involved.

> This question involves a non right-angled triangle and you must recognise that the sine rule can be applied to find *a* if the missing angle C can be determined. You will find a reminder of the sine rule in the **formula booklet**.

> You should clearly organise your answer including a short **explanation** at each stage so that the solution **is easy to follow**.

> You should state your final answer **clearly**, to an appropriate level of **accuracy** and including the **correct units**.

Now try this

Continue with the Worked example. Calculate the length of the second cable supporting the radio mast.

'Calculate' questions 2

Here are some examples of skills involved if answering 'calculate' questions where you need to find the number or amount of something based on information given in the question and, in addition to applying a mathematical method or formula, you also use additional subject knowledge. Some questions may require longer answers than others. The example below is of a fairly short answer.

Worked example

A steel tank has dimensions 1.5 m x 1.5 m x 1.5 m. The tank is used to store oil in a vehicle maintenance workshop.

A depth gauge indicates an oil level of 0.8 m inside the tank.

Assume the density of the oil is 860 kg/m³.

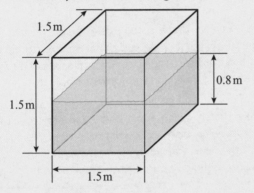

Calculate the hydrostatic thrust on one wall of the tank. **3 marks**

> The question has a scenario giving the **engineering context** for the question.

> You might find it useful to add your own labels or notes onto the diagram provided, or make a small **sketch** of your own to make things clear.

Sample response extract

From the formula booklet

$F = \rho g A x$

$\rho = 860$ kg/m³

$g = 9.81$ m/s²

$A = 0.8 \times 1.5 = 1.2$ m²

x = the height where the average pressure is found = $\frac{h}{2}$

$F = 860 \times 9.81 \times 1.2 \times \frac{0.8}{2}$

$F = 4049.568...$

$F = 4050$ N

> In this example, you must recognise that this question concerns the forces acting on an immersed plane surface. You will find the applicable formula in the Fluid principles section of the **formula booklet**.

> Your answer should **list** the variables used in the formula, **identify** each one and **establish** its value from the information given in the question. This will make your next steps far more straightforward.

> Once you have established values for each variable, these should be substituted into the formula. By **writing this out carefully** you will find it far easier to go back and spot any mistakes that you might have made at this stage.

> You should state your answer **clearly** using an appropriate level of accuracy and units. As a rule of thumb numerical answers are usually given to 2 d.p. or 3 s.f.

Now try this

A steel tank has dimensions 1.2 m x 1.2 m x 1.2 m. The tank is used to store kerosene to run a heating system. A depth gauge indicates the tank is half full. Assume the density of the kerosene is 810 kg/m³.

Calculate the hydrostatic thrust on one wall of the tank.

'Solve' questions

Here are some examples of skills involved when answering 'solve' questions.

Worked example

The tension in a drive belt at either side of a powered rotating pulley can be represented by the following equation:

$300 = 100e^{\pi\mu}$

where μ is the unknown coefficient of friction between the belt and pulley materials.

Solve the equation to find μ. Show evidence of the use of logarithms in your answer.   **3 marks**

Sample response extract

$300 = 100e^{\pi\mu}$

Take natural logs of both sides:

$\ln 300 = \ln 100e^{\pi\mu}$

Use the rule $\ln AB = \ln A + \ln B$:

$\ln 300 = \ln 100 + \ln e^{\pi\mu}$

Use the rule $\ln A^x = x \ln A$:

$\ln 300 = \ln 100 + \pi\mu \ln e$

Since $\ln e = 1$ then:

$\ln 300 = \ln 100 + \pi\mu$

Rearrange:

$\dfrac{(\ln 300 - \ln 100)}{\pi} = \mu$

$\mu = 0.350$

Check answer:

$100e^{(\pi \times 0.350)} = 300$

Now try this

The tension in a drive belt at either side of a powered rotating pulley can be represented by the following equation:

$350 = 100e^{2\pi\mu}$

where μ is the unknown coefficient of friction between the belt and pulley materials.

Solve the equation to find μ. Show evidence of the use of logarithms in your answer.

'Draw' questions

Here are some examples of skills involved when answering 'draw' questions.

Worked example

An electrical engineer is working on a system that combines two AC voltages with the same frequency into a combined output. The two AC voltage waveforms can be represented by:

$V_1 = 60 \sin(80\omega t)$

$V_2 = 90 \sin(80\omega t + \frac{\pi}{3})$

Draw a phasor diagram to represent $V_1 + V_2$ and find the magnitude and phase angle of the resultant phasor.

3 marks

Diagrams and graphs

In response to 'draw' questions, you might need to create a graph or diagram. This should be done accurately and to the appropriate scale. You might also need to label the diagram and provide clear annotations.

You can solve questions involving vectors or phasors **analytically** using trigonometry or **graphically** using an accurately drawn diagram. Here, the command verb '**draw**' tells you that the **graphical** approach is required.

Sample response extract

V_2 phase angle $\frac{\pi}{3}$ is equivalent to 60°.
Scale used: 1 cm to 10 V

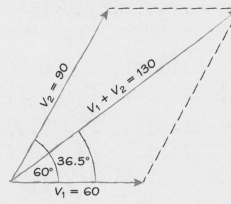

By measurement from diagram:

Magnitude of resultant phasor = 130 V

Phase angle (above horizontal) = 36.5°

To allow measurements to be taken from the diagram an **appropriate scale** needs to be chosen. This allows the drawing to fit comfortably in the space provided. In this example, a scale has been used of 1 cm to 10 V.

It is vital you use a **ruler** and **protractor** when drawing diagrams and that you work as accurately as possible. This example allows the use of a standard protractor by converting angles given in radians to degrees.

Take any measurements from the diagram carefully and accurately. You need to convert your measurements to the **units** required in the solution according to the scale you have chosen.

Although the question may not specifically ask that the diagram be labelled, if you correctly provide appropriate labels and annotation, it will help to make your **working clear**.

Now try this

An electrical engineer is working on a system that combines two AC voltages with the same frequency into a combined output.

The two AC voltage waveforms can be represented by:

$V_1 = 90 \sin(120\omega t)$

$V_2 = 80 \sin(120\omega t - \frac{\pi}{3})$

Draw a phasor diagram to represent $V_1 + V_2$ and find the magnitude and phase angle of the resultant phasor.

'Label' questions

Here are some examples of skills involved when answering 'label' questions.

Worked example

An electrical engineer is working on the designs for a transformer and is studying the B/H loop for a ferromagnetic material used in the transformer core.

On the graph, label the points at which the material reaches magnetic saturation and the point of coercivity.

2 marks

> When answering 'label' questions, you need to mark a particular feature or component on a diagram or sketch.

Sample response extract

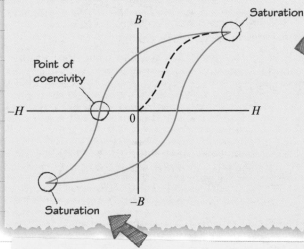

B/H diagram for ferromagnetic material

> You need to label all the characteristics asked for in the question.

> You need to clearly identify points or components with a leader line to the label.

Now try this

An engineer is working on the designs for a structural tie bar and is studying the results of a tensile test on the component.

On the graph, label the point of failure, UTS and limit of proportionality.

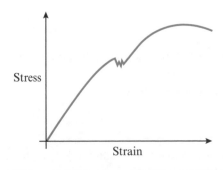

'Identify' questions

Here are some examples of skills involved when answering '**identify**' questions.

Worked example

A well-designed electrical transformer can achieve an efficiency of between 95% and 98.5%.

Identify two types of core losses that limit transformer efficiency.

2 marks

The first part of a question may establish the **engineering scenario** around which it is based.

Here, the question is testing your knowledge of transformer losses. Note that the question asks you to consider **core losses only**. You do not need to mention the ohmic losses caused by the resistance of the copper windings.

Sample response extract

Circulating eddy currents formed in parts of the iron core waste energy as heat.

Hysteresis losses are caused by the repeated reversal of magnetisation in the core.

When answering 'identify' questions, you need to recognise and recall the identity of particular features or processes. Keep your answers concise so that you use your time wisely for short and long answer questions.

Now try this

An electrical maintenance engineer is repairing a printed circuit board and is studying the corresponding circuit diagram.

Identify the following components that appear on the circuit diagram.

1

2

3

Synoptic questions 1

Here are some examples of skills involved if answering a synoptic question where you draw upon and link together knowledge from across Unit 1. This reflects the fact that you need a wide range of interconnected disciplines when tackling real-life engineering problems. A synoptic question may be divided into parts, as shown below and on page 89.

Worked example

On board a small ship electricity is provided by a diesel-powered generator. The diesel used to run the engine has a total energy content of 36 MJ/L and is consumed at a rate of 1.7 L/hr. The generator provides a 230 V electrical supply at 16 A. 8 kW of waste heat is generated by the diesel engine. This is transferred from the engine coolant system via a heat exchanger into glycol-filled radiators that are used to heat the living quarters. The glycol in the heating system has a specific heat capacity of 4.18 kJ/kg °C. The flow rate of glycol through the heat exchanger is 8.6 kg/s and the temperature difference maintained between the input and output pipes is +0.2 °C.

1 (a) Calculate the thermal efficiency of the heat transfer between the diesel engine and the glycol heating system.   7 marks

You will find that the scenario for a synoptic question can be long and contain lots of **information** that will be required to answer the questions that follow.

This example of part of the question requires that you make connections between **thermal efficiency**, **heat capacity** and **flow rates**. This **mix of topics** makes the question **synoptic**.

Sample response extract

$$\text{Thermal efficiency} = \frac{\text{Useful heat output}}{\text{Heat input}}$$

Heat input is 8 kW (waste from engine). This is equivalent to 8 kJ per second.

$$\frac{\text{Useful heat output}}{\text{Glycol heating}} \text{ is heat transferred to the system}$$

In 1 second 8.6 kg of glycol is heated by 0.2 °C in the heat exchanger.

From the formula booklet:

$Q = mc\Delta T$

$m = 8.6\,\text{kg}$

$c = 4.18\,\text{kJ/kg°C}$

$\Delta T = 0.2$

$Q = 8.6 \times 4.18 \times 0.2$

$Q = 7.1896\,\text{kJ}$

$$\text{Thermal efficiancy} = \frac{7.1896}{8} = 0.8987$$

<u>Required efficiency is 0.899 (to 3 s.f.)</u>

You should set out **calculate** questions like this one carefully to reduce the likelihood of confusion. Writing down the formula for how you will calculate the final answer is a good starting point. Then you should show how to obtain the information needed from the text of the question.

Now try this

On board a small ship electricity is provided by a diesel powered generator. The diesel used to run the engine has a total energy content of 33 MJ/l and is consumed at a rate of 1.8 l/hr. The generator provides a 110 V electrical supply at 32 A. 8.6 kW of waste heat is generated by the diesel engine. This is transferred from the engine coolant system via a heat exchanger into glycol filled radiators that are used to heat the living quarters. The glycol in the heating system has a specific heat capacity of 4.18 kJ/kg °C. The flow rate of glycol through the heat exchanger is 9.29 kg/s and the temperature difference maintained between the input and output pipes is +0.22 °C.

Calculate the thermal efficiency of the heat transfer between the diesel engine and the glycol heating system.

In this example of a synoptic question in two parts, the case study used for the Worked example above has been adjusted.

You can use the Worked example above to help you answer this example question.

Synoptic questions 2

This example of a synoptic question is the second of two parts and follows on from the case study and question on page 88.

Worked example

1 (b) Calculate the overall system efficiency.

7 marks

This example of a part of a synoptic question requires that you make connections between **electrical power**, the **energy content of fuel** and **flow rates**. This **mix of topics** makes the question synoptic.

Sample response extract

$$\text{Overall efficiency} = \frac{\text{Useful energy output}}{\text{Energy input}}$$

<u>Useful energy output</u> is the electrical power generated.

$P = VI$

$V = 2.30\,V$

$I = 16\,A$

$P = 230 \times 16 = 3680\,W$

A few notes or a sentence or two where necessary can help you to **organise** your thoughts and enable your working to be **easily followed**.

Heat <u>energy input</u> is the heat energy available in the fuel supplied to the diesel engine in the same time.

Energy of fuel used per hour = $36\,MJ/h \times 1.7\,l/h$

= $61.2\,MJ/h$ or 61.2×10^6 J/h

Converting tms to fuel used per second =

$$\frac{61.2 \times 10^6}{3600} = 17\,000\,J/s \text{ or } W$$

$$\text{Overall efficiency} = \frac{3680}{17000} = 0.216$$

<u>Required efficiency is 0.216.</u>

Try not to become overly reliant on the formulae booklet. You should aim to **remember** many of the more basic formulae, such as that used here to calculate electrical power.

You should allocate sufficient time to answer the final synoptic question so that you can take your time and work through each part **logically** and **methodically**.

You need to state your **final solution** clearly to an appropriate level of accuracy and include units. In this example, the final answer is an efficiency ratio and so has no units.

Now try this

With reference to the case study in Now try this on page 88, answer the second part of the synoptic question that follows.

Calculate the overall system efficiency.

You can use the Worked example above to help you answer this example question.

Using the formulae booklet

During your exam for Unit 1 you will have access to a booklet containing a list of formulae and physical constants. You must become familiar with this booklet, know what is included and what you will need to remember. This is shown on pages 91–95 and is included in the Sample Assessment Material for Unit 1 on the BTEC National Engineering section of the Pearson website. Always check the booklet in the latest Sample Assessment Materials to **make sure you are up to date**.

Formulae and constants

These include:

- static and direct current theory
- capacitance
- magnetism and electromagnetism
- single phase alternating current theory
- rules of indices
- rules of logarithms
- trigonometric rules
- volume and area of regular shapes
- the quadratic formula
- equations of linear motion with uniform acceleration
- stress and strain
- work, power, energy and forces
- gas laws
- angular parameters
- physical constants
- thermodynamic principles
- fluid principles.

Compound unit styles

Fundamental quantities such as mass, length or time have their own units. These are usually stated in their abbreviated form, which would be kg, m and s, respectively. Calculated quantities such as density, velocity or acceleration have what are known as compound units that are stated as a combination of fundamental units. You need to be familiar with two styles of expressing compound units. For example, in your Unit 1 exam and throughout this Revision Guide:

✓ The units for density are given as kg/m^3; elsewhere you may see this expressed as $kg\,m^{-3}$.

✓ The units for velocity are given as m/s; elsewhere you may see this expressed as $m\,s^{-1}$.

✓ The units for acceleration are given as m/s^2; elsewhere you may see this expressed as $m\,s^{-2}$.

The two styles can be used interchangeably and there is no right or wrong method, although you should be consistent.

Uses

1 Having the formulae booklet available means you can concentrate on applying your knowledge rather than remembering a large number of new formulae.

2 Use the booklet to check the formula you need. Don't get a question wrong just because you remembered a formula incorrectly.

3 If you're stuck, look through the list of formula; you may find it reminds you of the knowledge that you need.

Limitations

1 You won't be awarded any marks for remembering a formula when it is given in the booklet.

2 Many of the straightforward relationships, like trigonometric ratios, for example, are not included. You'll be expected to remember these.

3 When learning a topic do this alongside the formula booklet. Make a note of any formulae that are not included. You'll need to learn these.

4 Make sure you are familiar with the letters and symbols used in each of the formulae. The booklet does not define any of these terms. To be able to use the formulae you must learn how they are applied.

Now try this

Use the Formulae and constant booklet (pages 91–95) to find the formula used to calculate the following. Define all the terms used.

1 Voltage decay on capacitor discharge V_C.

2 The force between two electrostatically charged particles, F. (Hint: Coulomb's law.)

3 Hydrostatic thrust on an immersed plane surface, F.

Formulae and Constant

Static and Direct Current electricity theory

Formulae and constants

The formulae and constants booklet (pages 91–95) is in the Unit 1 Sample Assessment Material on the BTEC Nationals Engineering page of the Pearson website. Always check the website to ensure you are up to date.

Current $\qquad I = \frac{q}{t}$

Coulomb's law $\qquad F = \frac{q_1 q_2}{4\pi\varepsilon_0 r^2}$

Resistance $\qquad R = \frac{\rho l}{A}$

Resistance: temperature coefficient $\qquad \frac{\Delta R}{R_0} = \alpha \Delta T$

Ohm's Law $\qquad I = \frac{V}{R}$

Total for resistors in series $\qquad R_T = R_1 + R_2 + R_3$

Total for resistors in parallel $\qquad \frac{1}{R_T} = \frac{1}{R_1} + \frac{1}{R_2} + \frac{1}{R_3} \dots$

Power $\qquad P = IV, \;\; P = I^2 R, \;\; P = \frac{V^2}{R}$

Efficiency $\qquad E = \frac{P_{out}}{P_{in}}$

Kirchhoff's current law $\qquad I = I_1 + I_2 + I_3$

Kirchhoff's voltage law $\qquad V = V_1 + V_2 + V_3 \;\; \text{or} \;\; \sum PD = \sum IR$

Capacitance

Electric field strength $\qquad E = \frac{F}{q}$

Electric field strength: uniform electric fields $\;\; E = \frac{V}{d}$

Capacitance $\qquad C = \frac{\varepsilon A}{d}$

Time constant $\qquad \tau = RC$

Charged stored $\qquad Q = CV$

Energy stored in a capacitor $\qquad W = \frac{1}{2} CV^2$

Capacitors in series $\qquad \frac{1}{C_T} = \frac{1}{C_1} + \frac{1}{C_2} + \frac{1}{C_3} \dots$

Capacitors in parallel $\qquad C_T = C_1 + C_2 + C_3$

Voltage decay on capacitor discharge $\qquad v_c = Ve^{(-t/\tau)}$

Magnetism and electromagnetism

Magnetic flux density $\qquad B = \frac{\phi}{A}$

Magneto motive force $\qquad F_m = NI$

Magnetic field strength or magnetising force $\;\; H = \frac{NI}{l}$

Permeability	$\frac{B}{H} = \mu_0\mu_r$
Reluctance	$S = \frac{F_m}{\phi}$
Induced EMF	$E = BLv,\ E = -N\frac{d\phi}{dt} = -L\frac{dI}{dt}$
Energy stored in an inductor	$W = \frac{1}{2}LI^2$
Inductance of a coil	$L = \frac{N\phi}{I}$
Transformer equation	$\frac{V_1}{V_2} = \frac{N_1}{N_2}$

Single phase Alternating Current theory

Time period	$T = \frac{1}{f}$
Capacitive reactance	$X_C = \frac{1}{2\pi f C}$
Inductive reactance	$X_L = 2\pi f L$
Root mean square voltage	$RMS\ voltage = \frac{peak\ voltage}{\sqrt{2}}$

Total impedance of an inductor in series with a resistance $\qquad Z = \sqrt{X_L{}^2 + R^2}$

Total impedance of a capacitor in series with a resistance $\qquad Z = \sqrt{X_C{}^2 + R^2}$

Average waveform value average value $\qquad Average\ value = \frac{2}{\pi} \times maximum\ value$

Form factor of a waveform $\qquad Form\ factor = \frac{RMS\ value}{average\ value}$

Laws of Mathematics

Rules of indices

$a^m \times a^n = a^{(m+n)}$

$a^m \div a^n = a^{(m-n)}$

$(a^m)^n = a^{mn}$

Rules of logarithms

$\log AB = \log A + \log B$

$\log \frac{A}{B} = \log A - \log B$

$\log A^x = x \log A$

Trigonometric rules

Sine rule

$$\frac{a}{\sin A} = \frac{b}{\sin B} = \frac{c}{\sin C} \text{ or } \frac{\sin A}{a} = \frac{\sin B}{b} = \frac{\sin C}{c}$$

Cosine rule

$$a^2 = b^2 + c^2 - 2bc \cos A$$

Volume and area of regular shapes

length of an arc of a circle $s = r\theta$

area of a sector of a circle $A = \frac{1}{2} r^2 \theta$

volume of a cylinder $V = \pi r^2 h$

total surface area of a cylinder $TSA = 2\pi r h + 2\pi r^2$

volume of sphere $V = \frac{4}{3} \pi r^3$

surface area of a sphere $SA = 4\pi r^2$

volume of a cone $V = \frac{1}{3} \pi r^2 h$

curved surface area of cone $CSA = \pi r l$

Quadratic formula

To solve $ax^2 + bx + c = 0$, $a \neq 0$

$$x = \frac{-b \pm \sqrt{b^2 - 4ac}}{2a}$$

Equations of linear motion with uniform acceleration

$v = u + at$

$s = ut + \frac{1}{2} at^2$

$v^2 = u^2 + 2as$

$s = \frac{1}{2}(u + v)t$

Stress and strain

Direct stress $\qquad \sigma = \dfrac{F}{A}$

Direct strain $\qquad \varepsilon = \dfrac{\Delta L}{L}$

Shear stress $\qquad \tau = \dfrac{F}{A}$

Shear strain $\qquad \gamma = \dfrac{a}{b}$

Modulus of elasticity $\qquad E = \dfrac{\sigma}{\varepsilon}$

Modulus of rigidity $\qquad G = \dfrac{\tau}{\gamma}$

Work, power, energy and forces

Force $\qquad F = ma$

Resultant force $\qquad F_x = F\cos\theta,\ F_y = F\sin\theta$
(where θ is measured from the horizontal)

Mechanical work $\qquad W = Fs$

Force to overcome limiting friction $\qquad F = \mu N$

Gravitational potential energy $\qquad PE = mgh$

Kinetic energy $\qquad KE = \dfrac{1}{2}mv^2$

Gas laws

Boyle's Law $\qquad pV = \text{constant}$

Charles's Law $\qquad \dfrac{V}{T} = \text{constant}$

General gas equation $\qquad \dfrac{pV}{T} = \text{constant}$

Angular parameters

Centripetal acceleration $\qquad a = \omega^2 r = \dfrac{v^2}{r}$

Power $\qquad\qquad\qquad\qquad$ $P = T\omega$

Rotational Kinetic energy $KE = \dfrac{1}{2}I\omega^2$

Angular frequency $\qquad\qquad$ $\omega = 2\pi f$

Frequency $\qquad\qquad\qquad$ $f = \dfrac{1}{\text{time period}}$

2π radians $= 360°$

Physical constants

Acceleration due to gravity $\qquad\qquad$ $g = 9.81\,\text{m/s}^2$

Permittivity of free space $\qquad\qquad$ $\varepsilon_0 = 8.85 \times 10^{-12}\,\text{F/m}$

Permeability of free space $\qquad\qquad$ $\mu_0 = 4\pi \times 10^{-7}\,\text{H/m}$

Thermodynamic principles

Sensible heat $\qquad\qquad$ $Q = mc\Delta T$

Latent heat $\qquad\qquad$ $Q = ml$

Entropy and enthalpy $\qquad$ $H = U + pV$

Linear expansivity $\qquad$ $\Delta L = \alpha L \Delta T$

Fluid principles

Continuity of volumetric flow $\qquad\qquad\qquad\qquad$ $A_1 v_1 = A_2 v_2$

Continuity of mass flow $\qquad\qquad\qquad\qquad$ $\rho A_1 v_1 = \rho A_2 v_2$

Hydrostatic thrust on an immersed plane surface $\quad$ $F = \rho g A x$

Design triggers 1

Design triggers prompt the design of new or improved products or processes. They differ between **market-led** and **technology-led** companies, although in reality companies seldom use a single approach.

Market pull

Market-driven companies are constantly researching their **customers' needs** and what their **competitors** are offering. Their aim is to follow market trends and develop **new** or **improved products** to fill an identified need.

> The design and operation of online supermarket delivery services was triggered by market pull for a service.

Technology push

Technology-driven companies concentrate on **research** and **development** to **push the boundaries** of what is possible in consumer products. They tend not to have an established market and so are not guaranteed success. However, when they get it right it can be revolutionary.

> The inclusion of the first digital camera in a mobile phone was triggered by technology push in a product.

Demand

Demand is the number of products that could be sold into a particular market.

- **Market-led** companies try to measure demand through research.
- **Technology-led** companies hope to stimulate demand by launching innovative new products.

Profitability

Profitability is basically the **difference** between the **cost** of making a product and the **price** it can then be sold for. To maintain or increase profitability designers are often asked to redesign a product or service to reduce manufacturing or running costs.

Innovation

Innovation is the **key to the success** of both **technology-** and **market-led** companies. It is required to ensure that products remain **competitive**. The use of emerging technology or just approaching a problem in an unconventional way can transform the performance of an existing product or lead to new products with better performance for less cost.

> A series of innovations in bicycle design helped maintain the dominance of Great Britain's cycling squad at Rio 2016.

Now try this

Explain whether the following rely mostly on market pull or technology push to trigger design activities:

1 high street clothes retailer
2 Formula 1 team
3 mobile phone manufacturer.

Design triggers 2

New or improved designs are triggered in many interconnecting ways, including by market research, problems with a product or process, concerns about sustainability or the discovery of a risk.

Market research

Market research is essential for market-led companies to establish the **views of consumers** and find where **demand** for a product or service exists. The results of research are used to both trigger and inform the design of new products or prompt alterations to existing ones.

Focus groups

Product testing

Online surveys

Interviews

Who/what is involved in market research?

Consumer behaviour studies

Analysing competitor's products

Analysing market trends

Online communities

Performance issues

Performance issues with a product or process that results in **complaints** or **poor customer feedback** can also trigger a redesign.

For example:

- Computer hardware manufacturers continually strive to improve the performance of their products in order to cope with the demands of new software applications.
- Customer expectations for mobile phones to operate for long periods between charges has led to the rapid development of battery technology over the past 15 years.

Sustainability

Sometimes pressure from environmentally conscious consumers will drive a product redesign. Any increase in sustainability or decrease in **carbon footprint** will make a product more attractive to this large and growing group of customers.

For example, cars can be redesigned to:
- use energy efficient technologies to decrease fuel consumption
- reduce carbon emissions
- minimise the generation of other pollutants.

Product design also can be affected by legislation that is passed to limit pollution and protect the environment.

Designing out risk

Sometimes design changes are triggered in **response** to an incident, accident or the discovery of long-term health implications involved in using a product. For example, in the 1990s car engines were redesigned to run on lead-free petrol. This change was triggered when it was proven that exposure to lead in exhaust gases posed a potentially serious long-term health risk.

By 1992 all new cars had redesigned cylinder heads with hardened valve seats to allow them to run safely on unleaded petrol – a direct response to concerns about a serious health risk.

Now try this

You have been asked to design a child's toy.

Explain one potential risk that must be taken into account in each of the following areas:

1 the materials used
2 the form (or shape) of the toy
3 the function (or how it works).

Reducing energy

Designers are always under pressure to reduce the amount of energy used both in the design process and in the operation of a product during its life cycle.

Reduce energy usage in the design process

Taking advantage of the latest technology can help reduce the time, resources and energy taken to design new products.

Energy efficient design can be the direct result of:

- computer simulations
- 3D CAD modelling
- rapid prototyping
- online collaboration
- virtual design environments.

The Boeing 777 was the first airliner to be built and tested in a virtual design environment. A worldwide team created and assembled every component and simulated the operation of an entire virtual airliner before any real manufacturing began. This led to large energy savings during the design process.

Reduce energy usage during operation

Reducing the energy consumed when a product is in use can:

- reduce operating costs
- meet environmental obligations
- increase sales to environmentally conscious consumers.

As a designer you need to be aware of areas where energy savings might be possible. Some of these are outlined below.

Increase system efficiency – Use compatible components and subsystems that work together to minimise overall energy consumption; e.g. hybrid cars use a combination of a petrol engine and an electric motor to achieve high efficiencies with no loss of performance.

Increased component efficiency – Energy efficient components can make a big difference to the overall system efficiency; e.g. LEDs require significantly less electrical power than conventional filament lamps to achieve the same light output.

Reducing energy consumption

Reducing product mass – The heavier an object, the more energy is required to move it. Reducing mass makes products easier to move; e.g. F1 cars use carbon fibre composite technology to make them extremely light and so maximise the acceleration possible from the engine power available.

Reducing product dimensions – Commercial pressures for smaller products often lead to the miniaturisation of components. Smaller components tend to also use less power; e.g. miniature surface mount electronic components consume significantly less electrical power than an equivalent conventional through-hole component.

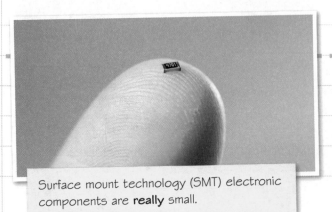

Surface mount technology (SMT) electronic components are **really** small.

Now try this

Explain two reasons why using a virtual design environment will save energy when developing a complex product, such as an aircraft.

Hybrids and energy recovery systems

Recent innovations in vehicle technology have led to an increase in the number of cars using hybrid combined power units and energy recovery systems. This makes them more efficient and reduces emissions.

Hybrid vehicles

Hybrid vehicles offset the disadvantages of using either electric motors or internal combustion engines on their own.

- With just an **electric motor** you could drive with zero local emissions but the distance travelled is limited by the size of your batteries, the time needed to recharge them and the limited number of charging points available. This is best suited to low speed, short journeys in towns and cities.

- With just an **internal combustion engine** you would contribute to local pollution levels but you can travel greater distances and it only takes a few minutes to refuel at one of many petrol stations to allow extended journeys. This best suits high speed, long journeys out of town and on motorways.

The challenge for designers and engineers is to **combine these technologies** in such a way that the driver gets a seamless driving experience and the best possible fuel efficiency from the vehicle.

In 1997 the Toyota Prius was the first mainstream commercially available hybrid car. The Prius is a parallel hybrid in which an electric motor and a conventional engine can be used either together or individually, depending on the driving conditions and power requirements.

Energy recovery systems (ERS)

The energy recovery system is one of the key aspects of **hybrid vehicles**.

- In a conventional car, when you need to slow down and operate the brakes, kinetic energy is converted to heat, which is wasted in the braking system as you slow down.

- In contrast, energy recovery systems slow the car by converting kinetic energy into electricity, which is then stored in on-board batteries until it's needed. Little, if any, energy is wasted as heat. On journeys in towns where heavy traffic often means stopping and starting repeatedly, energy savings from these systems can be significant.

- The impact of ERS has also been seen in motorsport. The latest ERS systems used in F1 allow engines to generate the same power as their predecessors but using 35% less fuel.

ERS systems in F1 motorsport are able to convert kinetic energy recovered from braking and heat energy recovered from the engine exhaust into electricity. This is stored for use to boost the available power when overtaking. Cutting edge technology developed in F1 normally trickles down into conventional high-cost then low-cost vehicles.

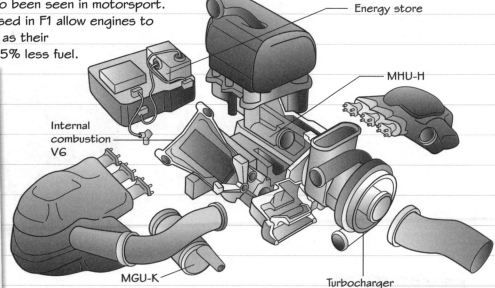

Energy store

MHU-H

Internal combustion V6

MGU-K

Turbocharger

Now try this

Describe two advantages of using a vehicle that uses hybrid technology.

Sustainability and cost over product life cycle

As a designer you have a responsibility to consider the **sustainability** and **costs** of designing and manufacturing a product. These should be considered over the **entire product life cycle**.

The product life cycle

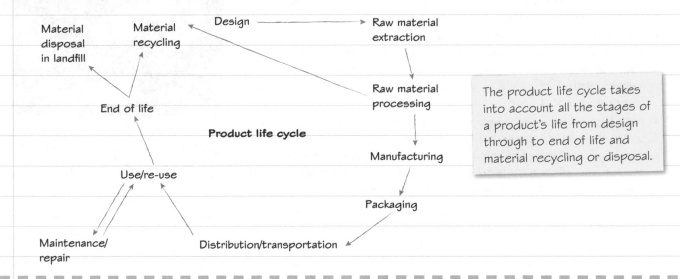

Design → Raw material extraction

Material disposal in landfill ← Material recycling

Raw material processing

End of life

Product life cycle

Manufacturing

Use/re-use

Packaging

Maintenance/repair   Distribution/transportation

The product life cycle takes into account all the stages of a product's life from design through to end of life and material recycling or disposal.

Cost reduction over product life cycle

A slightly more complex or expensive design or material specified at the design stage can be **offset** by reduced costs later in the product life cycle. For example, if you design a product so that the constituent materials can be separated easily, any additional design and manufacturing costs will be offset by lower recycling costs at the end of the product's life.

Sustainability of product life cycle

Manufacturers often have both a **legal** and **moral responsibility** to make their operations more sustainable. In addition, commercial pressure is also exerted by a more environmentally conscious public who demand products with greater sustainability. For example, most motor manufacturers refurbish and remanufacture used vehicle components and offer them for sale as replacement parts, instead of sending them for recycling.

The end of the road. VW aim to enable 95% of the materials and components in their vehicles to be recycled or re-manufactured for re-use at end of life.

Now try this

IKEA supply the majority of their furniture as flat packs for home assembly.

1   Describe one reason why this helps to reduce life cycle costs.
2   Describe one reason why this helps to increase sustainability.

High-value manufacturing and designing out risk

At the design stage it is important to consider the **potential risks** that the manufacture or use of your product might pose and whether **high-value manufacturing** techniques will be employed.

Designing out risk for employees

Good design should eliminate all types of risk to the employees who are manufacturing your products.

- Ensure that the **materials and processes** you specify are as safe to use and carry out as possible. For example, water-based paints or powder coating pose a reduced health risk compared to solvent-based paints and should be used where possible.
- Ensure that the risks of making mistakes during **manufacturing and assembly** are minimised. For example, using asymmetric mounting holes will ensure parts can only be assembled in the correct orientation. This principle is known in Japan as 'Poke Yoke', which roughly translates as making a process fool proof.

Designing out risk for customers

Good design will eliminate any **health and safety** risk to stakeholders, customers and the end users of your products. This will involve examining the **materials** used, the **form** and the **function** of the product.

For example:

- Food and drink containers should only ever use food-safe materials that won't contaminate their contents (for example, drinks bottles manufactured from PET thermoplastic polymer).
- Sharp edges or corners on any product should be rounded off to prevent accidental injury.
- The mechanism of a folding chair must avoid forming a trapping hazard in which fingers might be caught during operation.

Commodity and high-value manufacturing

Commodity manufacturing is characterised by low value and high volume. It uses low technology traditional processes and production methods. Much commodity manufacturing has been outsourced to low-cost countries and emerging economies.

In contrast, **high-value manufacturing** is characterised by lower volume, high complexity and high value. Manufacturing is often dependent on innovative new processes, manufacturing techniques and materials driven by investment in research and development. This can be a very challenging design environment but in fully developed industrial economies it is vital for continued growth of domestic manufacturing. For example, the manufacture of Airbus wings in the UK has all the characteristics of high-value manufacturing.

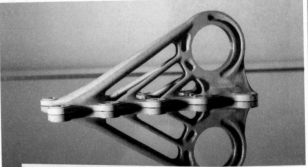

Additive manufacturing of titanium brackets used by the aircraft industry has led to reduced resources, lower weight and increased strength. It has also increased the cost of the product, but this is offset easily by savings in fuel.

Now try this

The standard UK mains plug has been in use since the 1940s and has several safety features built into its design.

1 Describe two features of a plug that have been designed to reduce the risk of harm to those using it.

2 Is the production of UK mains plugs an example of commodity or high-value manufacturing?

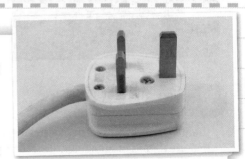

Systems, equipment and interfaces

Large engineering systems can be broken down into individual pieces of **interconnected** equipment. These work together to perform the function of the system as a whole.

Systems

System is a term used to describe often large and complex assemblies or machines that depend on the coordinated function of interconnected smaller subsystems and individual pieces of equipment. Examples include an aeroplane, an automated production line or a car.

Equipment

Equipment is a term used to describe the individual interconnected elements that make up a larger system. For example, a robot arm is just one piece of equipment that might be found on an automated production line. Several of these working in conjunction with other equipment, sensors and control technology might be required in a complete system.

Interfaces

The types of connections or interfaces between equipment in a system depend on the required complexity, flexibility, intelligence and the technology employed. Often a range of technologies can be used in the same system. For instance, a car integrates mechanical, electrical, hydraulic, pneumatic and microcontroller systems.

Mechanical – Combinations of levers, linkages, cables, pulleys and gears can provide the required level of interconnection and control in mechanical systems. Such systems are generally robust and reliable but lack flexibility.

Electrical – Combinations of relays, connectors and switches wired together into electrical circuits can provide control in electrical systems. Such systems are generally robust and reliable but lack flexibility.

Equipment interfaces

Hydraulic/Pneumatic – Combinations of various valve types connected using pipework into hydraulic/ pneumatic circuits can provide the required level of control in hydraulic/pneumatic systems. Such systems are generally robust and reliable but lack flexibility.

A centrifugal speed regulator on a steam traction engine – these used entirely mechanical interfaces and control technology.

Microcontroller/software – The advent of microcontrollers and software-controlled systems of electromechanical relays, valves and actuators have revolutionised what is possible when interfacing equipment: different types of technology can be combined in the same system far more simply. Electronic sensor technology, combined with intelligent control software, allows system behaviour to be changed in real time in response to a range of external factors, and software-based controllers provide almost infinite system flexibility.

In cars, mechanical carburettors have been largely replaced by electronic fuel injection and engine management systems that provide superior performance and fuel economy. These can precisely alter the mix of fuel and air entering the engine in response to a wide range of parameters monitored by the system. They can also be reconfigured to alter the engine performance using a laptop.

Now try this

Create a simplified sketch of a small system within your field of study. Include details of equipment-level components, such as motors or sensors, and the technology used to connect and control them.

 For example, sketch the system used to operate anti-lock brakes on a motor vehicle.

System compromises

Invariably some trade off or compromises must be made when integrating all the equipment required within a system. Sometimes it is necessary to accept the shortcomings of some elements in order to achieve the best overall system performance.

Systems integration

The issue of systems integration and how this can often lead to compromise is best illustrated using a familiar example, in this case, a car.

Cooling – Engines generate waste heat that must be dissipated to prevent overheating. Cars therefore incorporate additional cooling equipment in the form of radiators to overcome this problem. However, the waste heat from the engine can be turned into an opportunity by redirecting it to heaters in the passenger compartment when required.

Bonding – Physically supporting and joining the different equipment in a car is done almost exclusively by spot welding and the use of fasteners, such as screws, clips, and nuts and bolts. Ideally the body shell and panels would be permanently welded together to provide maximum rigidity; however, removable access panels are often required to allow the repair and servicing of other equipment-level components.

System integration compromises in a car

Location (for optimum equipment performance) – To allow a radiator to function efficiently and provide effective engine cooling it should be placed at the front of the vehicle in a constant flow of cool air. This is true even for mid-engine vehicles where the engine is behind the driver. This usually means fitting long coolant pipework back to the engine. In F1, a radiator at the front of the car would ruin its aerodynamics and, as a compromise, the radiator is mounted further back and ducting or scoops are used to divert air through it to provide cooling.

Electrical/electronic compatibility – Electrical systems in standard motor vehicles run on 12 V DC. The original purpose of a battery was to provide a means of starting the engine. Modern cars feature a huge number of other electrical and electronic equipment, from radios to windscreen wipers. To allow simple equipment integration all these also run on a 12 V supply even if that means including devices within the equipment to reduce this voltage to make it compatible with their electronic systems.

Centre of gravity – To ensure stability when cornering, the centre of gravity in a car should be as low and central as possible. This is achieved at system level by the careful positioning of equipment-level elements. Therefore, when heavy equipment such as the engine is located at the front of a car it is counterbalanced by the fuel tank at the rear. (In motorsport all the equipment, including the driver, is positioned as low as possible.)

F1 cars have a mid-engine and are extremely low to the ground in order to optimise the position of the centre of gravity for fast cornering. The radiators are positioned in the side pods at either side of the driver.

Now try this

Equipment-level waste heat generated by a car engine provides an opportunity at system level to make use of this elsewhere.

Describe a different example from within your field of study where waste generated at equipment level is put to use elsewhere in the system.

For example, consider how heat from an aircraft engine can be used to de-ice the wings.

Equipment specifications and cost effectiveness

Each piece of equipment in a large, complex system is seldom designed or manufactured by a single company or in a single facility. Specifying equipment requirements accurately is therefore essential.

Equipment product design specifications (PDS)

When specifying equipment for use in large systems you must ensure equipment product design specifications are detailed and accurate. These must consider a range of equipment and system-related issues.

Shortcomings absorbed at system level – Can sensitive pieces of equipment be positioned together and benefit from shared system-level protection from heat, cold, dust and vibration?

Electromagnetic compatibility (EMC) – Will the equipment generate electromagnetic interference that might affect other electrical or electronic equipment? Can this be minimised or eliminated? Is electromagnetic shielding required or can sensitive equipment be repositioned or shielded instead?

Specification considerations

Cooling – How much waste heat will be generated by the equipment? Can it be used elsewhere in the system or must it be dissipated using its own or system-level cooling equipment?

Mass – What is the maximum mass of the equipment that will allow the overall system mass to be achieved?

Cost effective production

A systems approach to design and manufacturing can have several advantages, including lowering production cost.

- **Off the shelf equipment (proprietary)** – Where possible use off-the-shelf standard equipment, as it avoids equipment design and manufacturing costs (**capital outlay, set up** and **tooling costs**).

- **Specialists and centres of excellence** – Sometimes specialist equipment must be designed and manufactured. Investing in centres of excellence and concentrating specialist expertise, manufacturing capability and tooling in one facility for each specialism makes the overall process more efficient.

- **Concurrent development** – This means that each specialist piece of equipment or subsystem can be worked on by different teams at the same time, minimising overall lead times and cost.

A systems approach in action

Airbus have specialist centres of excellence, designing and building aeroplane equipment and subsystems throughout Europe.

Now try this

Aerospace is one industry that takes a systems approach to design and manufacturing.

Describe another product or industry that adopts a similar approach.

For example, describe how the automotive industry takes a systems approach to design and manufacturing.

Mechanical properties

As a designer you will need to understand the mechanical properties of a range of materials and how they compare. Mechanical properties describe how a material is affected by the application of different forces.

Yield point – The stress after which any further strain or extension will be permanent (e.g. when a bolt that has been stretched will not return to its original length). This is really a measure of the maximum useful tensile stress.

Yield point

Tensile testing apparatus being used to experimentally determine the UTS and yield point of a steel sample.

Ultimate tensile stress (UTS) – The maximum tensile stress a material can withstand before failure.

Compressive strength – The maximum compressive strength a material can withstand before failure; e.g cast iron has the ability to withstand large downward loads typically supporting machine tools or bridges.

Shear strength – The maximum shear stress a material can withstand before failure.

Fatigue limit – The maximum level of stress that can be applied cyclically without the risk of fatigue crack formation and eventual failure. (Components with sharp corners and repeated vibrations suffer from fatigue.)

Mechanical properties of materials

Hardness – The ability of a material to resist indentation, scratching and wear (e.g. a file demonstrates hardness by cutting a softer material).

Elastic modulus – The stiffness of a material when undergoing elastic deformation.

Toughness – The ability of a material to resist fracture under shock loading.

Malleability – The ability of a material to be deformed permanently using compressive forces. (Any item that is forged is malleable.)

Ductility – The ability of a material to be stretched permanently using tensile forces.

Now try this

Select two mechanical properties of high speed steel (HSS) and explain why these make HSS a suitable material for use in drill bits.

Physical and thermal properties

As a designer you need to understand the physical and thermal properties of a range of materials and how they compare.

Physical properties of materials

Boiling point (°C) – Defines the temperature at which a liquid changes phase into a gas (or boils).

Melting point (°C) – Defines the temperature at which a solid changes phase into a liquid (or melts).

Density (kg/m^3) – Measures the compactness of a material by measuring the mass present per unit volume.

Appearance – Qualitative characteristics including colour, texture and other aesthetic properties.

Physical properties

High melting points

Ceramic materials tend to have extremely high melting points. This ceramic-lined crucible is able to contain molten steel at around 1500°C.

Latent heat of vaporisation (kJ/kg) – The amount of energy absorbed when one kilogram of a material changes phase from liquid to gas.

Latent heat of fusion (kJ/kg) – The amount of heat energy absorbed when one kilogram of a material changes phase from solid to liquid. When the material is changing its phase (state) the temperature and pressure remain static.

Thermal properties of materials

Specific heat capacity (J/kg°C) – Measures the amount of heat energy required to raise the temperature of one kilogram of material by one degree.

Thermal conductivity (W/mK) – Measures a material's ability to conduct heat. Heat travels from hot to cold.

Thermal properties

Coefficient of linear expansion (m/m°C) – Measures the linear expansion of a material with each degree rise in temperature.

Now try this

Describe one **physical** and one **thermal** property of the material used for the skin of a spacecraft that will help ensure that it survives the heat of re-entry into the Earth's atmosphere.

Electrical and magnetic properties

As a designer you will require an understanding of the electrical and magnetic properties of a range of materials and how they compare.

Electrical properties

Resistivity (Ωm) – Measures the degree to which a material resists or opposes the free flow of electrons. Current flow depends on the movement of electrons.

Conductivity (1/Ωm) – The inverse of resistivity, conductivity measures the degree to which a material permits the free flow of electrons.

> **Electrical properties**

Temperature coefficient of resistivity (1/°C) – Measures the change in resistivity of a material with each degree rise in temperature.

Magnetic properties

Permeability (H/m) – Measures the degree of magnetisation a material undergoes when inside a magnetic field.

Reluctance (1/H) – Measures the resistance a material has to the passage of a magnetic field through it.

> **Magnetic properties**

Ferromagnetism – Certain metals, such as iron, nickel or cobalt and their alloys, can be described as being ferromagnetic. Ferromagnetic materials are strongly attracted by a magnetic field and readily become magnets themselves when exposed to an external magnetic field.

Electromagnets

The operation of an electromagnet depends on the magnetisation of its ferromagnetic core by a magnetic field generated by a coil in which an electric current flows.

Now try this

You have been asked to design a shield to protect some sensitive electronics from the effects of stray magnetic fields. Shielding works on the principle of providing a low resistance path for the magnetic field to pass through, directing it away from the area being protected.

Describe the magnetic property of the material used for the shield that determines its effectiveness.

Advanced materials

Material science is continually advancing and as a design engineer you should keep up to date with advanced, new and novel materials and their potential applications.

Biomaterials

Biomaterials can be metals, polymers or ceramics. Their only common characteristic is that they are **compatible** for use inside the body without disrupting or adversely effecting biological systems.

Applications of biomaterials include:

- joint replacements
- bone fixings
- dental implants
- heart valves
- stents.

Smart alloys

Smart alloys are a group of specialist metal alloys that exhibit novel characteristics such as:

- **superelasticity** – these alloys can undergo large strains and still spring back to their original shape when conventional metal alloys would have been permanently deformed.
- **shape memory** – these alloys can appear to have been plastically deformed and bent into shape but upon heating will spring back to their original shape.

Nitinol

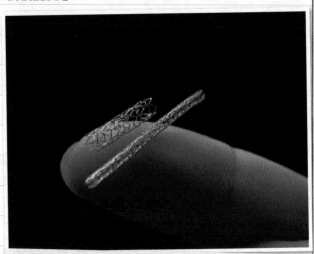

Nitinol, an alloy of nickel and titanium, is both **biocompatible** and a **smart** alloy that exhibits shape memory characteristics. One medical application for nitinol is in arterial stents used to widen restricted arteries. The mesh stent is compressed and placed inside the artery. The stent is then warmed in situ, which causes it to expand to its original shape, and push against the artery walls, allowing improved blood flow.

Nanoengineered materials

When materials are engineered into very small particles or structures on the nano scale (1–100×10^{-9} m) they often exhibit a range of novel properties not present in other forms of the material.

Nanoengineered materials have found applications in a range of consumer products, and new nano materials such as graphene (a form of carbon), are the subject of much academic research.

Applications of nanoengineered materials include:

- sunscreen – uses nanoparticles to absorb UV rays
- carbon fibre structures – can be stiffened using silica nanoparticles
- lithium-ion batteries – use electrodes based on nanoparticle technology
- flexible solar cells – made possible by nanoparticle technology
- Self-cleaning glass – used in construction and relies on a coating of nanoparticles.

Now try this

You have been asked to design some high-end spectacle frames for a high street retailer.

You could go on to identify applications of advanced materials used in your own field of study.

From the options above select a type of advanced material that would be suitable in this application and justify your choice.

Surface treatments and coatings

The components you design will often require the application of a surface treatment to protect them from corrosion. Of course, coatings can also enhance the aesthetic appeal of a finished product.

Anodising
- Commonly applied to aluminium and titanium, which form inert and hard-wearing oxide layers spontaneously in air.
- Anodising uses an electrochemical process to increase the thickness of this layer and provide greater protection.
- Anodised finishes are porous and are often dyed in bright colours before being sealed.

Electroplating
- Chromium plating is a long-established method of protecting brass taps or steel car bumpers from the effects of corrosion.
- It also gives a highly reflective, polished surface finish.

Surface treatments and coatings

Galvanising
- Galvanising is widely used on steel products for use outdoors and involves applying a zinc coating to the product.
- This forms a physical barrier to prevent the ingress of water, which can lead to rusting.
- It also electrochemically protects the steel, providing long-term maintenance-free corrosion resistance.

Powder coating
- Powder coating is often used as an alternative to wet paint, which provides similar levels of protection and colour choice.
- It is generally considered to be less environmentally harmful, as it avoids the use of solvents.
- It works by applying electrostatically charged thermoplastic powder to metal components and then baking the component in an oven to fuse the powder into a uniform coating.

Painting
- Painting provides a protective layer to prevent the ingress of water, which leads to corrosion in metals and rot in wooden materials.
- Just as importantly, paint is available in many colours and is used to provide enhanced aesthetic appeal to a wide range of products.

These galvanised gates will stay maintenance-free for many years.

Now try this

You are restoring an old Land Rover and find the chassis is heavily corroded and needs to be replaced.

Describe which of the above surface protection and coating(s) would provide the best protection for the replacement chassis to maximise its service life.

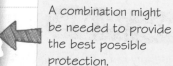

A combination might be needed to provide the best possible protection.

Lubrication

The components you design will often require the use of appropriate lubrication to protect them from wearing prematurely.

Lubrication

To prevent premature wear, overheating and inefficient running most mechanical systems require lubrication.

Reduce friction
The lubricant film reduces friction between surfaces.

Reduce wear
The lubricant film prevents direct contact between surfaces and so mechanical surface erosion is reduced.

Purposes of lubrication

Reduce heat build-up
Circulating lubricants carry heat away from the components being protected.

Protect against corrosion
Lubricants that are liquid or grease-based provide a barrier to moisture, which prevents corrosion.

Remove debris
Circulating lubricants can wash and carry away dirt or debris from moving components. The dirt is then captured in a filter.

Fluid film lubrication

Fluid film is an effective **lubrication regimen**, with examples found commonly in the moving parts of motor vehicles. It depends on a constant film of lubricant supporting the full load between two surfaces to prevent any physical contact between them.

There are two common ways to achieve this:

- **Hydrostatic lubrication**, where the lubricant is pumped under pressure between the bearing surfaces to prevent it from being squeezed out.

- **Hydrodynamic lubrication**, which relies on the relative movement of the surfaces to pressurise and maintain an effective lubricant film.

Types of lubricant

1. **Liquid** – this may be circulating, such as in a car engine, or it may be sprayed or dripped on. In the case of circulating lubricants and systems where it is applied directly, the lubricant and filter should be changed regularly.

2. **Grease-based** – this lubricant may be packed around moving parts. For some components, the grease may be sealed in and designed to last the life of the part, but for others reapplication may be needed, and a grease gun can be used to apply fresh grease as required.

3. **Solid** – these are commonly based on graphite. They have an advantage over liquid lubricants because they can operate at high temperatures without degrading.

Now try this

Identify three reasons why effective lubrication is necessary in a car engine.

Modes of failure of materials

In order to avoid or protect against component failure you must be aware of the common failure mechanisms that you might encounter and the operating conditions that cause them.

Brittle fracture
- Generally occurs in high-strength materials with low toughness and poor ductility; e.g. cast iron, high carbon steel, glass, most ceramics.
- Caused by overloading in tension.
- There is little or no plastic deformation or necking (a reduction in diameter of the material as it elongates) evident prior to brittle fracture and so it can occur quickly and without any warning signs.

Ductile fracture
- Generally occurs in materials with high toughness and good ductility; e.g. copper, wrought iron, mild steel.
- Caused by overloading in tension.
- Considerable plastic deformation, permanent elongation and necking can be observed before failure.

Creep failure
- Creep is a phenomenon where a constant tensile stress well below the yield point of the material leads to plastic deformation and elongation over extended periods of time.
- If left unchecked this will lead to eventual component failure.
- The rate of creep is elevated at high temperatures.

Modes of failure

Corrosion
- A familiar form of corrosion is rust on iron and steel.
- All forms of corrosion involve chemical processes that attack the material itself.
- This gradually affects the material's appearance and eventually weakens its mechanical properties.

Fatigue failure
- Fatigue is a phenomenon encountered in components subject to cyclic loading (repeated loading and unloading) well below their normal yield strength.
- A tiny material flaw or scratch can be enough to initiate a fracture that grows slowly with each cycle.
- Over time cracks can grow sufficiently to weaken the material and cause component failure.
- Localised stress concentration at sharp corners or other features on a component can accelerate the fatigue process.

Fracture surface showing the characteristic beach marks of fatigue crack propagation (the gradual growth of a small crack in the material which enlarges slightly with every loading cycle).

Now try this

The turbine blades in jet aircraft engines operate at stress levels below their material's yield point but at temperatures close to its melting point.

Identify the mode of failure most likely to affect the metal alloy used in jet engine turbine blades when operating in these conditions.

Mechanical motion

As a designer of mechanical systems you need to appreciate the four basic types of motion that you can employ.

Linear
Possibly the simplest type of motion, where movement is along a **straight line**. An example of linear motion could be observed in the hydraulic cylinder operating a bearing press.

Rotary
Rotation describes motion along a **circular path** about some central point or centre of rotation.

> **The four types of mechanical motion**

Oscillating
Oscillating motion can be described as a form of **rotary** motion where direction is **reversed** periodically and a repetitive **swinging** motion is established. A swinging clock pendulum is a good example of oscillating motion.

Reciprocating
Reciprocating motion describes a form of linear motion where direction is **reversed** periodically and a repetitive **up and down or back and forth** cycle is established. The two stages of a reciprocating cycle are called **strokes**.

Rotary and reciprocating motion

Piston

Rotating craft shank

> The crank shaft in an engine provides a good example of **rotary** motion. Also shown here is the piston, which demonstrates **reciprocating** motion.

Now try this

From within your field of study identify at least one example for each of the four types of mechanical motion.

Mechanical linkages

In most mechanical systems, you will need to link the motion of one element of a system to another. You may also need to convert one type of motion into another, change its speed, direction or the force it can exert.

Mechanical linkages

The are many forms of mechanical linkage that you will need to be familiar with.

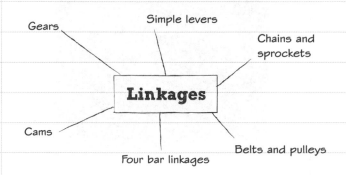

Gears

Simple levers

Chains and sprockets

Linkages

Cams

Four bar linkages

Belts and pulleys

Cams

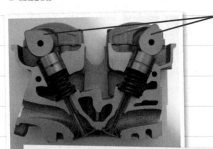

Cams

This photo shows the cutaway of an engine with twin overhead camshafts. The rotating cams that open and close the engine valves can be seen clearly.

Mechanical advantage

One of the key aspects of linkages is their ability to take a small force acting over a large distance and convert it into a larger force acting over a proportionally smaller distance. This is the principle by which a lever works.

This is known as **mechanical advantage** (MA), which is defined as the ratio between the output force or load (F_l) and the input force or effort (F_e) where:

$$MA = \frac{F_l}{F_e}$$

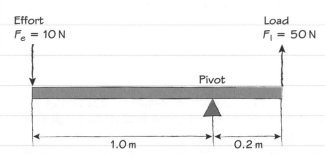

Effort
$F_e = 10\,N$

Load
$F_l = 50\,N$

Pivot

1.0 m   0.2 m

A simple lever, where $MA = 50 \div 10 = 5$

Linkages in nature

Structures comparable to some types of mechanical linkage can be found in nature.

For example, short linear muscle contractions in your biceps use a basic linkage to provide almost 180° of rotary motion at your elbow.

A Mantis shrimp uses a system of levers to enable its raptorial claws to be accelerated rapidly and pack an enormous punch.

Now try this

Research and sketch two examples of the mechanical linkages mentioned in the spider diagram at the top of the page. One application of cams has been given for you.

For one example, you could sketch the linkage that operates a windscreen wiper. Research and sketching is useful revision of the skills you need for your set task.

Power sources

When designing a product it's important to select the most appropriate way to power it. You should always aim to use the most **clean**, **efficient** and **sustainable** power source possible.

Mechanical power

Mechanical power drove the industrial revolution and was being harnessed long before the development or widespread use of electricity.

Sources of mechanical power include:

- internal combustion engine
- steam engine
- steam turbine
- gas turbine
- fly wheel
- clockwork mechanism
- falling weight
- windmill
- waterwheel.

Mechanical and electrical power are inextricably linked as the vast majority of power generation technologies convert mechanical power; e.g. from a steam turbine in a power station, into electricity.

The BayGen Freeplay radio

This was the first radio that did not need an electricity supply or batteries. Instead, it used an internal clockwork generator. The hand-wound clockwork mechanism provided the mechanical power to drive a small electric generator, which in turn provided the electricity needed for the radio's electronics.

Electrical power

Electricity provides a source of power and runs the control technologies that are at the heart of engineering and technology. Although the majority of the electricity we consume is generated in **power stations** there are **other sources of electrical power**, including:

- battery
- hydrogen fuel cell
- photovoltaic cells
- wind turbines
- hydroelectric generators
- thermoelectric generator (teg)
- power station mains electricity (gas/coal/nuclear).

Energy from nature – fuels

Some forms of power generation, such as solar or wind, extract energy from **naturally occurring** and **renewable** sources. However, the majority rely on the consumption of some kind of fuel. **Fossil fuels,** which contribute to climate change, are still by far the most common. As a designer it's your responsibility to **steer away** from fossil fuels and minimise their use throughout a product's life cycle.

Types of naturally occurring fuels include:

- hydrogen
- coal
- oil
- biomass
- gas
- petrol
- alcohol
- diesel
- biodiesel.

Now try this

Identify three different types of power source encountered in your area of study and provide an example of where each is used.

Controlling power transmission

Sensors, **controllers** and **actuators** are the three main elements of an engineering system. They monitor the condition of a system, decide whether some action is necessary and take the required physical action when required.

Sensors

Sensors monitor the **condition** of the system and send that information to the controller.

Types of sensors include:

- pressure sensor
- temperature sensor
- speed
- displacement/position sensor
- load cells (measuring forces)
- limit switches
- light sensors.

Controllers

Controllers monitor **information** provided by sensors and decide when to take **action** by operating an **actuator**. In engineering you will encounter purely electromechanical, pneumatic and hydraulic control systems. More commonly, electronic controllers now tend to be used in most applications.

Types of controllers include:

- pneumatic circuits
- hydraulic circuits
- electromechanical systems
- embedded microcontrollers
- computers running condition-monitoring and control software
- programmable logic controllers (plcs), commonly found in industrial machinery.

An engineering system example

Sensor

Modern intelligent heating systems use sensors in different areas to monitor the temperature around your home.

Controller

Signals from the temperature sensors are monitored by a central controller and if any fall below a preset level the system responds.

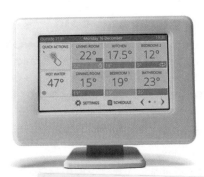

An intelligent heating system controller, which monitors the temperature throughout your home.

Actuator

The signal from the controller activates a pump in the heating system, which circulates hot water to the radiators in the area where additional warmth is required.

Actuators

Actuators carry out **physical movements** in response to pneumatic, hydraulic, mechanical or electrical signals from the controller.

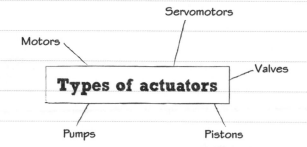

Motors — Servomotors — Valves — **Types of actuators** — Pumps — Pistons

Now try this

Identify an engineering system related to your area of study. Identify the types of sensors, controllers and actuators used in its operation.

Processing metals

The basic processes for manufacturing metal components fall into three categories: **casting**, **forming** and **machining**.

Casting

Casting involves pouring **liquid molten** material into a **mould**. You can use a number of techniques, depending on the complexity, physical size and the required surface finish on the component.

Metal casting techniques include:

- die casting (hot chamber)
- die casting (cold chamber)
- sand casting
- investment casting.

Investment casting

This process uses a lost **wax** process to create seamless ceramic **moulds**. These produce intricate, dimensionally accurate castings with good surface finish in high-melting point materials.

Here, stainless steel is being poured into moulds designed to cast several components at once.

Forming (hot and cold)

Forming uses **mechanical force** to **reshape** a solid billet or sheet of material. Some metals can be shaped cold, whereas others need to be heated to increase malleability and/or ductility prior to forming.

Hot-forming processes include:

- drop forging
- press forging
- rolling
- extrusion.

Cold-forming processes

These processes include:

- cold rolling
- cold (or impact) extrusion
- coining
- wire drawing
- spinning
- shallow/deep drawing
- press forming.

Press forming is used to shape sheet material. Here, block-and-blade tooling is being used on a press to form straight, uniform bends in sheet steel components.

Machining

Unlike casting and forming, **machining processes** remove excess material from a workpiece to **form** the required component shape.

Machining processes include:

- drilling
- turning (using a lathe)
- milling.

CNC lathes

In most industrial manufacturing applications **computer-controlled** machine tools, like this CNC **lathe**, are used to increase speed, quality and repeatability.

Now try this

Aluminium bracket

A colleague in your company has designed an aluminium bracket (as shown) that will be required in large numbers.

Select a suitable process or processes for manufacturing this bracket and justify your selection.

Powder metallurgy and additive manufacturing

The use of powder metallurgy is not new and similar processes have been used for centuries in the manufacture of ceramics. However, modern materials and techniques, including the emergence of additive manufacturing, have led to an increase in its use.

Powder metallurgy

Powder metallurgy is a mass manufacturing process in which fine metal powder is **compacted** together in **moulds**. At this stage the components are in their **green state** and are very delicate, having only the strength required for handling during production.

The green components are then sintered, which involves heating in a controlled atmosphere until the metal particles fuse together permanently. The amount of **compaction** during **moulding**, **sintering temperature** and **time** all influence the amount of shrinkage that occurs and the **density** of the finished component.

Components made by powder metallurgy have no need for additional machining to remove sprues or flash. They have good dimensional accuracy and the materials used can be tailored to match specific applications, including mixing metallic and non-metallic materials.

Additive manufacturing

In additive manufacturing green components are not moulded **but built up in layers** using one of a range of 3D printing techniques.

Indirect 3D printing

Indirect metal printing uses 3D inkjet printers to deposit a binder material on specific areas of a layer of fine metal powder. When this dries a further layer of powder is added followed by more binder deposited by the printer. The component is built up in this way, layer by layer, until it is complete. It is then removed from the 3D printer in its green state and sintered in a conventional way.

Direct 3D printing

Direct Metal Laser Sintering (DMSL) is capable of **fusing** together a finished component directly without the need to create a green stage. In practice, the materials suitable for this process are limited to those with relatively low sintering temperatures, such as aluminium.

DMLS components emerging from unused powder. The use of additive manufacturing in this way is ideally suited for low-volume production or prototypes as it eliminates the need to invest in mould tooling.

Now try this

Give two advantages of using powder metallurgy over an alternative conventional casting process, such as die casting.

Links To revise casting techniques, see page 116.

Joining and assembly

You need to be familiar with a range of **permanent** and **non-permanent** methods of joining and assembling metal components.

Permanent

Many joining and assembly methods are permanent and, as such, cannot be easily undone to repair or replace components.

Permanent methods of joining and assembly include:

- **Adhesives** – epoxy resins can be used to permenanently join dissimilar materials together; e.g. metal and polymer components.
- **Welding** – there are a number of welding systems used to permanently fuse similar metals together; e.g. electrical resistance welding, oxy-acetylene, metal inert gas (MIG) and tungsten inert gas (TIG).
- **Spot welding** – this is a form of electrical resistance welding used to join sheet metal parts.
- **Brazing** – this relies on the adhesion of a brass alloy to the surface of steel components to form a joint. Brazing is carried out at a lower temperature than welding so avoids the effects of heat distortion on thin-walled components.

Non-permanent

Non-permanent joining and assembly methods are used where components need to be repaired or replaced, or when they are easier or more convenient than permanent methods.

Non-permanent methods of joining and assembly include:

- **Self-tapping screws** – frequently used to join thin sheet metal components. The specially designed and hardened screw is able to tap its own thread as it is screwed into the sheets.
- **Blind threaded inserts** – these are fixed into sheet materials to provide a permanently held threaded hole to accept a machine screw. This provides a neater solution than a self-tapping screw and greater strength, whilst also allowing for repeated removal and replacement.
- **Machine screws** – have threads along their entire length and a head to allow tightening in one of several styles (e.g. slotted, hexagon, Philips, Pozi-drive, socket). Screws are fitted through a clearance hole in one component and into a threaded hole in a second, to join them together.
- **Nuts and bolts** – bolts are similar to screws but are generally not threaded along their entire length. They are meant to be used in conjunction with a nut and washer to join two components through plain clearance holes.

Computer controlled robotic **spot welding** is often used to join car bodies.

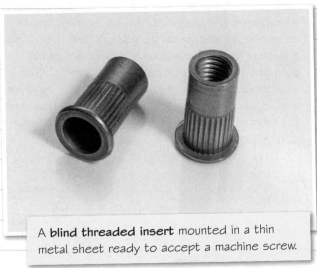

A **blind threaded insert** mounted in a thin metal sheet ready to accept a machine screw.

Now try this

Most bicycle frames are constructed from several sections of thin steel tubing.

Select a suitable process for joining the frame components together and justify your selection.

Processing polymers

Polymers are an incredibly **versatile** range of **synthetic materials** that can be shaped into anything from forks to fridge doors using a wide range of manufacturing processes.

Casting – Large thermoset polymer resin components, such as kitchen work surfaces and bathroom surrounds, are cast in large moulds.

Additive manufacturing – Techniques include stereolithography, 3D printing or fused deposition modelling. Commonly used in rapid prototyping creating components directly from a 3D model.

Injection moulding – Heated thermoplastic is injected into a two-part metal mould, rapidly cooled and then ejected from the mould tooling. High volume thermoplastic polymer components with complex shapes are usually injection moulded.

Polymer manufacturing processes

Compression moulding – Powdered resin is compressed in a mould to the required shape. The resin is cured by the application of heat and pressure to make thermoset polymer components with complex shapes.

Extrusion – Long, uniform lengths with a uniform cross-section are formed when heated thermoplastic is forced or extruded through a shaped die plate.

Thermoforming – Thermoforming processes, such as vacuum forming, stretch heated thermoplastic sheet over basic moulds to form components with uniformly thin walls.

Complex thermoplastic **injection moulded** component.

Thermoplastics sheet materials, such as cast acrylic, are made by **casting** heated liquid polymer between sheets of glass.

Now try this

Identify a suitable process for the high volume production of the component shown and justify your selection.

Processing ceramics

Crystalline ceramics are a range of hard, brittle materials with extremely high melting points often used as abrasives or coating in cutting tools.

Selective laser sintering (SLS) – An **additive** manufacturing technique where a computer-guided laser is used to thermally fuse thermoplastic polymer-coated ceramic particles together into the required shape. This is ideal for low-volume production runs or prototypes.

Slip casting – Ceramic powder mixed with a liquid to form a slurry is able to flow into a mould. The liquid is drawn off and the component dried to leave the required shape.

Injection moulding – Ceramic powder mixed with a thermoplastic polymer can be injection moulded to form complex shapes in high volume.

Processing ceramics

Extrusion – Ceramic powder held in a binder material can be extruded through a shaped die plate to form long lengths with uniform cross-sectional area.

Loose powder being brushed away from a completed **SLS** component.

Dry axial or isostatic pressing – Dry ceramic powder is compressed in a mould to the required shape.

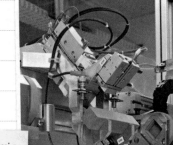

Extrusion of ceramic honeycomb profile.

Firing (or sintering)

All the processes outlined above provide the shape of the required component. However, until the component is fired or sintered at high temperatures it does not have its characteristic **strength**. Before firing, any binder materials used need to be removed by gentle heating or by dissolving them using a solvent.

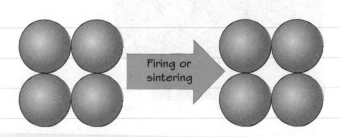

Firing or sintering

During firing, atoms diffuse between adjacent particles, causing them to permanently fuse together.

Now try this

An aerospace company needs to manufacture a prototype of a rocket nozzle from a high-melting point ceramic material.

Select a suitable process for the manufacture of ceramic prototype components and justify your selection.

Processing fibre-reinforced composites

Composite materials are made up of two or more different materials which, when used together, provide greater strength and stiffness than either material can provide on its own. In engineering applications you will commonly encounter **glass fibre-reinforced plastic** (GFRP) and **carbon fibre-reinforced plastic** (CFRP).

GFRP lay up

Lay up involves building up **layers** of oriented glass fibre matting and polyester resin inside a mould to the required thickness. You usually use lay up for load-bearing GFRP components.

GFRP spray up

Spray up involves **spraying** a mix of chopped glass fibre strands and polyester resin into a basic mould to build up the required thickness. You usually use spray up for non-load bearing GFRP components.

A GFRP component being manufactured using spray up.

CFRP moulding

Moulding uses **prepreg** CFRP sheets (carbon fibre matting impregnated with epoxy resin).

To mould CFRP components:

- Cut prepreg sheets to the required size.
- Layer up sheets into a mould to the required thickness.
- Bag and seal the mould.
- Apply a vacuum to compress the layered sheets and eliminate any gaps.
- Cure the component in an autoclave at elevated temperature and pressure.
- Remove the component from the mould and trim.

CFRP-automated fibre or tow placement

A **tow** is a narrow ribbon or tape of carbon fibre material. You can manufacture large or complex components by applying layers of fibres or tows using a **computer-controlled** automated process. Each tow is applied, trimmed to length, heated and pressed into place by a robotic applicator.

Automated fibre placement robot in action.

Now try this

Select a fibre-reinforced composite material and suitable process to manufacture a composite garage door. Justify the choice of material and process.

You could go on to identify applications of fibre-reinforced composites used in your own field of study.

Effects of processing

The mechanical properties of materials can be significantly affected by how they are **processed** during manufacture or subsequent heat treatment.

Processing metals

Heat treatment can have a significant effect on metallic materials. It is widely used to change mechanical properties by altering the **crystalline grain structure** of the metal.

- Metals that have been fully **annealed** (re-crystalised by heating and gradual cooling) have large, irregular grains and tend to be relatively soft, making machining processes easier.

- After machining, **hardening** heat-treatment processes are often used to encourage the formation of smaller grains and alternative crystal structures, which tend to provide greater hardness and strength.

- **Alloying elements**, too, can be used to strengthen metals, usually by straining the normal crystal structures within the metal and so changing their properties.

Forging

Forging takes a metal billet (a partly finished round or square bar) and uses impact or compression forces to form the required shape.

During the forging process long, uniform and tightly packed grains are formed, which enhance strength and toughness. The grain flow can be seen in these cross-sections of forged components.

Processing ceramics

Engineering ceramic components can be shaped using a number of techniques prior to sintering. The parameters used in these processes will affect the properties of the components produced.

Ceramic process parameters include:

- particle size
- compaction density
- sintering time
- sintering temperature
- heating rates
- cooling rates.

Processing composites

The ratio in which the different materials are used in fibre-based composites will affect their mechanical properties.

- In general, increased strength and stiffness is achieved by increasing the **proportion** of the fibre reinforcement phase of the material.

- Another important consideration is the **orientation** in which fibres are used. To provide extra strength in one particular direction, fibre matting can be replaced with a similar number of single fibres all acting in the direction of the principal loading stress.

Processing polymers

The main manufacturing method used for polymer components is injection moulding. The **parameters** used during moulding affect the quality and properties of the components produced.

Injection moulding process parameters include:

- injection pressure
- injection temperature
- mould temperature
- clamp pressure
- injection time
- cooling time.

Now try this

A CFRP-composite component manufactured using woven carbon fibre matting has failed at one particular point in its structure that is subject to a large directional load.

Suggest a way to reinforce this area, making the most efficient use of materials and without significantly increasing the size and weight of the component.

Scales of production

The **volume** and **variety** of products that need to be produced usually dictates the scale of production used. This, in turn, defines the **approach** taken to manufacturing.

Scales of production

 One-off

A single unique item made to order. This might be the skilled work of a small number of craftsmen to produce a piece of bespoke furniture or the combined efforts of hundreds of workers on a large, complex project like an oil rig.

 Small batch

Small manufacturing runs of products in the tens or possibly hundreds. This might be several identical hand-crafted chairs that make up a dining set or the manufacture of highly specialised F1 engines required to complete a race season.

 Large batch

Large manufacturing runs in the hundreds or possibly thousands. This might be a run of bespoke lighting fixtures for an entire office block.

 Mass

Very large manufacturing runs in the tens of thousands or more. This might be the manufacture of a single product, such as a mobile phone over 12 months or more, until manufacturing changes to a new model.

 Continuous

Indefinite manufacturing runs where a constant demand must be met over several years. This might be the manufacture of disposable or consumable products, such as machine screws, nails or paperclips, which are consumed in their millions.

Fixings such as screws are manufactured continuously in their millions.

Characteristics of the different scales of production

Scale of production / Characteristic	One-off	Small batch	Large batch	Mass	Continuous
Variety	Very high	High	Low/Medium	Low	One
Volume	One	Low	Medium	High	Very high
Unit cost	High	Medium	Low	Low	Very low
Tools and equipment	General	Specialised	Specialised and some dedicated	Specialised and some dedicated	Dedicated
Initial investment	Low	Medium	High	High	High
Production efficiency	Low	Medium	High	High	Very high
Labour type	Skilled	Skilled and semi-skilled	Semi-skilled and unskilled	Semi-skilled and unskilled	Unskilled

Now try this

Identify one example of an appropriate product for each of the five scales of production.

Customers

There are two types of customer who might trigger the design process: **internal** and **external**.

External customer

These are customers who use your goods and services but are **not** part of your organisation.

For example:

- another manufacturing company that uses your components in its products
- a person or organisation who uses your finished products
- a wholesaler who stocks your products and supplies them to other companies
- a retailer who stocks your products and supplies them directly to end-users.

Internal customer

These are customers **within** your organisation to whom you supply goods and services.

For example:

- another department in your company that your department supplies with components or sub-assemblies, e.g. at Jaguar Land Rover the main vehicle production line is the customer for the engine manufacturing plant
- a commercial department within your business, such as sales or marketing, which might ask you to design a new product to fill a gap in the market.

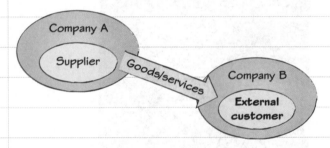

External customers – the supplier and customer belong to different companies or organisations.

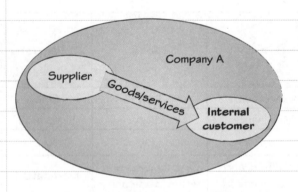

Internal customers – the supplier and customer are both in the same company.

Now try this

Identify example scenarios in which the design process might be driven by:

(a) an internal customer

(b) an external customer.

Product and service requirements

During the design process you must take into account a range of product and service requirements to ensure your product meets the client's expectations. The following questions should be answered to establish product and service requirements. The requirements of a mobile phone have been used as an example.

Performance specifications – How well must the product complete its functions? What factors affect the product's key performance characteristics? For example, processor speed, charge time, battery life, memory, display resolution?

Operating standards – Which existing products or infrastructure must the product be compatible with? For example, charger supply voltage, mobile digital transmission standards GSM, 3G, 4G or 5G, region-specific operating frequencies, bluetooth?

Manufacturing quantity – What volume of products will be required? How will this affect the scale of production used? How much investment in specialised tools and equipment will be required? For example, mobile phones are mass manufactured in high volume using dedicated specialist tooling and equipment with high set-up costs.

What might the requirements be for a mobile phone?

Reliability – What will be the user's expectation of product life? What operating conditions can be expected? What are the consequences of failure? Is maintenance required?
For example, Consumers expect to keep a mobile phone for up to two years during which it must be 100% reliable.

Product support – Will you be expected to provide support on the use of the product? What level of support might be needed? Do you need to stock spare parts and consumables? What instruction and documentation should be provided? Will software updates and hardware upgrades be available? For example, mobile phone manufacturers handle repairs through their retailers, spare parts are available and are generally fitted professionally, no user maintenance is required, software updates are issued periodically.

Life cycle – Where did the materials to make the phone come from and what happens to them at the end of the product life? For example, mobile phones tend to use new materials, including some rare metals.
At the end of life a proportion of these can be recovered through recycling.

Usability – How easy is the product to use? Are the controls intuitive and straight forward? Will user training be required? For example, consumers expect to be able to use a mobile phone straight out of the box. All controls and functions must be easily accessible and intuitive to operate.

Anthropometrics and ergonomics – How will people interact with the product? Does the size, weight and shape of the product correspond with the size, strength and shape of an average user. Are a range of sizes required? For example, a mobile phone must be held in one hand, buttons must be sufficiently separated to be operated by a finger tip, weight must be minimised to prevent fatigue, and position of buttons and controls must make them comfortable to operate.

Now try this

Choose a product or service you are familiar with and state one example of a product or service requirement for each of the headings used above.

Product design specification (PDS) 1

It is essential that a designer understands the needs of their customer. A **PDS** is used to help the **designer** and **customer** to **agree** exactly what is required. Below are **eight** areas that a PDS should specify.

 Cost: The target manufacturing cost per product. This is usually a commercial decision based on a proportion of the expected selling price.

 Quantity: The manufacturing quantity. This will dictate the scale of production and so the investment required to set up facilities and the manufacturing processes used.

Product design specification (PDS)

 Maintenance: The level of maintenance required. This will define minimum service intervals and maintenance schedules.

 Finish: The type of finish needed. This might be decorative, coloured, reflective, wear resistant, corrosion resistant or require a combination of characteristics.

 Sustainability: Opportunities for using sustainable materials and processes. Where possible, manufacturing processes should be energy efficient and pollution free. Waste materials should be reused or recycled.

Aesthetics: Aesthetic requirements. Some products are purely functional whereas others have to blend in with a specific decorative style, period or design movement.

 Weight: The maximum or minimum weight of the finished product. This will heavily influence the choice of materials.

 Materials: The mechanical and physical characteristics of the materials required. This will include any restrictions on the type of materials that can be specified.

Links To revise further criteria for the PDS, see page 127.

Carbon footprint

The PDS could also contain information about how carbon emissions generated during the manufacture and use of the product should be minimised.

Carbon footprint should be considered over the whole **product life cycle** and should include raw material extraction, manufacture, transportation and energy consumption in use.

Now try this

Identify a product that you are familiar with from your line of study. For each of the eight headings above, write down one item that could have been specified in the PDS for that product.

Product design specification (PDS) 2

Below are **ten** areas that a PDS should specify in order to inform the designer and meet customer needs.

 Links These follow on from the criteria for the PDS outlined on page 126.

 Safety: As a designer, have you considered your legal and moral responsibility to make sure your products and the processes and materials used to manufacture them are safe?

 Testing: Do provisions have to be built in for user calibration and testing? Will specialist test kits or tools be needed? Will the product be able to report errors to the user?

 Reliability: What will be the end user's expectation? Is the product disposable, for short-term or long-term use? Is the product safety critical? Is regular maintenance required?

Ergonomics/Anthropometrics: How do people interact with the product? Is it comfortable to use and compatible with the shape, size and strength of the person using it? Will you need to provide a range of size options?

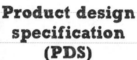 **Product design specification (PDS)**

 Manufacturing processes: Is the budget available to invest in new techniques, processes and equipment? Can the design use existing equipment when it is manufactured? Can processing enhance the properties of the materials used?

 Usability: Is the user experience positive? Is the product easy to use effectively? Are the controls easily understood and operated?

The function of some buttons has become universally understood.

 Manufacturing constraints: Does the required capacity for production already exist? Do you have the skills and expertise in the processes required? How long will it take to set up suitable production facilities?

Competition: How must the design compare with a competitor's product? What is your USP or performance advantage? What useful innovations adopted by the competition can you make use of or build upon?

A typical high-volume manufacturing facility.

 Manufacturing facilities: Are the manufacturing facilities required for your product already in place? Does the design need to take into account the capabilities of existing facilities?

 Market: How well developed is the market for your product? What kind of customers are you trying to appeal to? How big is the market? How many of the products are you likely to sell?

Now try this

Identify a product that you are familiar with from your line of study. For headings 1 to 5 from the diagram above, write down one item that could have been specified in the PDS for that product.

Commercial protection

You can **protect** your innovative ideas, designs or technologies from being copied or used without your permission by various forms of **commercial protection**.

Patents

Patents are concerned with **product function** (how something works).

A **patent**:

- gives you the sole right to use a novel idea, process or technical innovation that you have invented
- prevents others from exploiting or copying your ideas for a period of up to 20 years
- enables you to sell or licence your ideas to other people for a fee.

James Dyson, inventor of the bagless vacuum cleaner, holds over 560 patents to protect his ideas.

Design Registration ®

Design Registrations are concerned with **product form** (how something looks).

Design Registration:

- protects the individual character, shape or appearance of your product
- prevents others from using the same or similar designs for their products
- enables consumers to recognise your products easily.

One of the world's iconic registered designs. Do you recognise it?

Copyright ©

Copyright can be used to protect your **original work**. We usually think of copyright on books, films or computer games. It can also apply to commercial software packages, engineering drawings, images and web pages.

Copyright:

- prevents others from reproducing, distributing or using your work
- enables you to sell or licence your work for a fee.

Trademarks™

Trademarks protect the use of a distinctive graphic, word or logo to **represent your company** or **organisation**.

Trademarks:

- can be registered to your company or organisation
- prevents competitors from using it on their products.

Some familiar trademarks.

Now try this

What kind of commercial protection would be most appropriate in the following examples?

(a) A photograph you have taken.

(b) A traditionally styled tin opener that you manufacture.

(c) The name used for a chain of shops you have opened.

Legislation and standards

When designing a product you must remember that if products do not comply with relevant legislation, national or international standards, then they may not be saleable.

Limitations

Legislation can sometimes **limit** what is permissible when designing a product. It can also prompt the need for design changes.

For example, in 2014, European environmental protection regulations banned the import or production of vacuum cleaners above 1600W. This led to the withdrawal of a range of existing products, including the Miele S5212 2200W vacuum cleaner.

Opportunities

Changes in legislation can also offer **opportunities**.

For example, after 2014 Miele needed to offer more efficient high-performance vacuum cleaners and beat their competition to market. They developed technology to overcome the power limitations and at the same time maintain overall performance with their EcoLine models, which use only 800W. They are now a market leader in low-power, high-performance cleaners.

Legislation

Certain responsibilities and minimum safety standards are set out in law. **Legal requirements** differ between countries. For example:

- Health and Safety at Work Act 1974 applies in the UK.
- Compliance to General Product Safety Regulations (GPSR) is required throughout the European Economic Area.

Approved codes of practice (ACOP)

These are issued in the UK by Health and Safety Executive (HSE). ACOPs give **practical advice** on how best to comply with health and safety regulations. Non-adherence to ACOPs can be used as evidence against you in a prosecution.

ACOPs include:

- control of substances hazardous to health
- personal protective equipment at work
- controlling noise at work.

Standards

These cover a vast array of products and services. Designers need to be **familiar** with their standards and to **stay abreast** of new guidelines that may be issued.

Standards:

- represent agreed best practice
- are designed to be used vountarily
- give consumers confidence in the quality of your product if there is certification to a recognised standard
- are both national and international (see logos opposite).

European Standards Organisation (CEN, CENELEC, ETSI).

National certification

German Institute for Standardisation (DIN).

bsi.

DIN

British Standard (BSI).

International certification

cen **CENELEC**

American National Standards Institute (ANSI).

ANSI
American National Standards Institute

ISO

International Standards Organisation (ISO).

Now try this

An update to the ISO standard on the safe design of bicycle frames means that a bicycle manufacturer must redesign one of their products if it is to maintain its ISO approval.

State an example of a positive opportunity this might represent for the manufacturer.

Environmental and safety constraints

Environmental and safety constraints, often dictated by legislation, standards and codes of practice, will also influence the design process. As a designer you must consider the safety and environmental impact of a **product** and the **processes** used to make it.

Environmental constraints

These include:

- **Sustainability** – use renewable resources when available and use non-renewables efficiently and responsibly.
- **Carbon footprint** – minimise the amount of CO_2 released into the atmosphere through the use of fossil fuels or other industrial processes.
- **Product life cyle** – consider environmental impact at all stages, from initial design to end of life.

Product Health and Safety constraints

These include:

- **Form** – there must be no sharp edges, corners or other features likely to cause injury.
- **Function** – the operation of the product must pose no risk to the user and you should provide instructions for its safe use.
- **Material** – the materials must be non-toxic and pose no risk to the user or when recycled or disposed of at end of the product's life.

Process Health and Safety constraints

Material processing – the materials specified should pose no significant safety risk to those extracting, processing or working with them during manufacturing.

Manufacturing methods – the working environment and manufacturing methods used during production should pose no significant safety risk to those involved. This should also include the processes used to manufacture bought-in components, especially those from low-cost manufacturing countries where worker safety is sometimes less of a priority.

Considering the whole product life cycle

Designers have a responsibility to minimise the energy and resources consumed by a product and the safety of consumers and workers at every stage in its life cycle. That means you must be aware of the different stages that a product will go through during its life.

 Links Look back to page 100 and the diagram showing the product life cycle.

Creating aluminium drinks cans

Aluminium generally starts its life as the aluminium ore bauxite and this raw material is usually extracted from large open-cast mines.

The first stage:

- Open-cast mining causes local environmental damage and pollution.
- In poorly regulated countries there are real safety concerns for mine workers.
- Large amounts of electricity are required to process the ore and extract aluminium in a usable form.
- Overall, the extraction process has a large carbon footprint.

Now try this

Aluminium drinks cans are used in their millions every day.

Continuing the example of considering the life of an aluminium drinks can, describe the potential impact on the environment at each stage of the product life cycle.

Security constraints

Security constraints for both **products** and **processes** can also influence the design process. With our expanding reliance on computers, and the drive for 'always on' wireless connectivity, keeping systems secure is a growing concern.

Product security

The security of products is no longer limited to physical anti-theft precautions (although these still play a part) and include:

- **physical anti-theft systems**; e.g. locks, alarms, security marking
- **preventing product counterfeiting**; e.g. bank notes
- **securing data** held in microcontroller-enabled, computers or smart devices; e.g. protecting personal data held on bank cards, mobile phones, laptops or tablets
- **securing remotely controlled or monitored systems** from unauthorised access, cloning or hijacking; e.g. protecting vulnerable devices such as wireless security systems or cameras, keyless cars and remotely operated equipment such as drones
- **securing the control of autonomous systems** such as driverless cars
- **securing information** on new product developments and prototypes from the prying eyes of the public, journalists and competitors.

Autonomous driverless cars might one day be commonplace on our roads. These are controlled by complex computer software and incorporate cutting-edge technologies and wireless connectivity. It is very important that these systems are made secure from external interference that might otherwise cause accidents, serious injury or death.

Process security

Sometimes the security surrounding **processes** is just as important as for products.

- Commercial processes often involve steps that are kept secret to help the manufacturer maintain an **advantage** over their competitors.
- Many industries use potentially **dangerous** processes or materials that must be kept secure; e.g. radioisotopes used in medical equipment and nuclear power generation, toxic or poisonous chemicals.

Industrial secrets at Rolls Royce protect details of their process to manufacture lightweight hollow turbine fan blades from titanium by inflating them with an inert gas whilst they are red hot.

The UK nuclear industry has its own armed police force, the Civil Nuclear Constabulary, to ensure dangerous materials are protected and secure.

Now try this

Select a product or process from your area of study and describe at least one security constraint that affects its operation or design.

Marketing

Products need to have a clear competitive edge, a trusted reputation and effective marketing in order to become successful.

Unique selling point (USP) – What makes your product stand out from the competition? What can your product do that others can't?

Benefits of the design – What are the benefits of your product in comparison to the competition? How do price, performance and style compare?

Marketing goals: considerations

Obsolescence – How fast is technology and fashion moving? How quickly will a product become outdated? How often will consumers expect improved and upgraded performance and/or style? How do you develop brand loyalty and so repeat custom?

Obsolescence

In some markets consumers have accepted the cost of rapid obsolescence. To keep up to date with the latest models you would need to have bought six incarnations of the iPhone between 2007 and 2015 at a cost of around £4000!

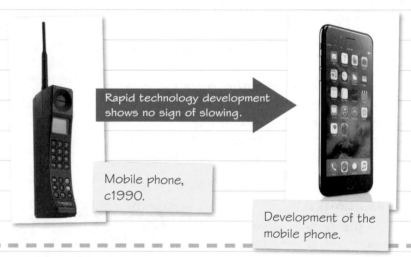

Rapid technology development shows no sign of slowing.

Mobile phone, c1990.

Development of the mobile phone.

USP

Marketing teams use USPs to make their products more attractive. It might be a low price, but it is more likely to be a feature or performance that outstrips the competition.

USPs should be:

- **true** (Don't make false claims!)
- **wanted/needed** by customers
- **protected** – If a design can be copied and you have not protected your intellectual property, competitors can copy your innovations.

Dyson's USP

The USP for the first Dyson vacuum cleaner was that it did not use a dust bag. That proved extremely attractive to consumers. However, the product would not have been a commercial success if overall performance, form and function had not also been as good, if not better, than its conventional competitors.

Now try this

Choose a familiar product that you purchased yourself. Consider the following:

1 Does it have a USP?

2 Give one reason why it stands out from the competition.

3 How long do you expect it to last before it is replaced?

Form and functionality

When you are designing a product, it is essential that you consider what it looks like (form) and what it is required to do (function).

Form

Form defines how a product should **look**. You can explore this by considering an outdoor camping chair.

Colour – Muted colours that don't clash with the outdoor environment. Use of more than one colour will add interest. Colour of painted legs should complement the colour of the fabrics used.

Shape – Curved and comfortable looking (this will partly be dictated by required function). Compact and neat when folded.

Form

Texture – Fabrics should be non-slip, soft and warm with padding in high-wear areas.

Style/Aesthetic appeal – Must be attractively styled to attract consumers and complement the colour and style of other outdoor and camping equipment, such as tents and clothing. The product must retain its appearance even after an extended period outdoors.

Function

Function defines exactly what a product is intended **to do**. For example, the function of the camping chair should ensure it:

- is strong enough for use by an adult
- is comfortable during extended use
- is stable on uneven ground
- is collapsible to allow easy transportation and storage
- is weatherproof
- supports back and arms
- incorporates a drinks holder.

Design emphasising form

Sometimes form, style and aesthetic appeal are just as important as function.

Design emphasising function

Sometimes form is less important and almost entirely dictated by function.

Alessi corkscrew.

Traditional corkscrew.

Decorative glass chandelier.

Industrial explosion proof bulkhead lighting.

Now try this

Look closely at the Alessi corkscrew above. Suggest two functional and two form requirements that were considered when designing this product.

Product performance

As well as form and functionality a range of other performance objectives and technical considerations must be considered when designing a product.

Choice of materials and components

It is essential that you select appropriate materials and components to suit the performance requirements of your product. You need to consider:

- **characteristics affecting material choice**, including strength, stiffness, fatigue strength, thermal conductivity, thermal expansion, density, ductility, corrosion resistance, electrical conductivity, cost
- **characteristics affecting component choice**, including power consumption, size, weight, compatibility with other equipment, cost, reliability, maintenance needs.

Sustainability

Sustainability should be at the forefront of your mind when designing new products.

Consider the product's impact on the environment and its carbon footprint:

- Can recycled materials be used?
- Are non-renewables used efficiently?
- Can energy consumption be reduced?
- Will it be straightforward to recycle at end of life?
- Can worn parts be reconditioned or remanufactured for reuse?

Interactions with other areas/components

You must ensure that a product is able to operate effectively in a given working environment and interface with other components where necessary.

Consider the product's:

- ergonomics
- anthropometrics
- physical size
- space requirements
- access for servicing or repair
- electrical compatibility
- heating/cooling needs
- resilience to environmental factors; e.g. vibration, damp.

Likelihood of failure/wear

You must also consider the consequences of product failure or wear. These must fall within required user expectations of reliability and service life.

Failure and wear considerations include:

- redundancy
- ease of repair
- consequences of failure
- expected service life
- required level of maintenance
- expected reliability.

The concept of redundancy is widely applied in safety critical applications, such as in the oil industry. It involves the use of two or more independent methods to control and carry out safety critical tasks so that any single instance of equipment failure will not cause catastrophic failure of the system overall.

Product redundancy – a critical issue

Now try this

Explain the two most important characteristics you should consider when selecting a material for the moulded case of an electrical plug.

Manufacturing processes and requirements

It is important to keep in mind the tools and techniques that will be required to manufacture the products that you design. Without an understanding of a range of **tools**, **processes** and **techniques** you will not be able to make efficient use of them.

Manufacturing processes

There is a huge range of processes used in manufacturing.

Metals – die casting, sand casting, investment casting, forging, extrusion, rolling, drawing, spinning, press forming, turning, milling, drilling, powder metallurgy, additive manufacturing.

Polymers – injection moulding, casting, thermoforming, extrusion, compression moulding, additive manufacturing.

Manufacturing processes

Composites – lay up, spray up, moulding, automatic tow/fibre placement.

Ceramics – slip casting, injection moulding, extruson, dry pressing, sintering (firing).

Assembly processes

An assembly line is classed as a **manufacturing process** where workstations add components or pre-assembled units to the semifinished product. An assembly process can be **manual** or **automated** (robot technology) or a **combination**. Assembly processes are used in the automotive, electrical / electronic sectors, from small fast-moving assemblies, such as a PCB for a phone, to slow moving, such as the aeronautical assembly lines.

An automated assembly line.

Manufacturing requirements

Other factors will also have a significant impact on the manufacturing processes selected, including:

- availability of skilled labour
- labour costs
- required machine time
- customer lead time
- tooling required
- CNC programming
- equipment
- factory space
- jigs and fixtures required
- machine capacity
- existing expertise and competencies.

Now try this

Recommend a suitable manufacturing process for this aluminium component and outline its manufacturing requirements.

Aluminium automotive engine component.

Manufacturing needs

As a designer you need to consider a range of **manufacturing issues** during the design process, which might influence the processes you select.

Quality indicators

Quality indicators are the key dimensional tolerances or other characteristics of a component that must be within specified requirements for it to be fit for purpose.

Common quality indicators include:

- dimensional accuracy
- surface finish
- appearance (witness marks, flash, ejector pin marks, sprues, burrs)
- damage (cracks, scratches, dents).

Surface finish process capability and selection chart

Ra μm	50	25	12.5	6.3	3.2	1.6	0.8	0.4	0.2	0.1
Machining processes										
Sawing		▨	▨	▨	▨	▨				
Shaping		▨	▨	▨	▨	▨	▨			
Drilling			▨	▨	▨	▨	▨	▨		
Milling		▨	▨	▨	▨	▨	▨	▨	▨	
Turning		▨	▨	▨	▨	▨	▨	▨	▨	
Broaching			▨	▨	▨	▨	▨			
Reaming			▨	▨	▨	▨	▨			

Charts like this are used to help designers select appropriate machining processes for use in manufacturing. Some processes might not be capable of providing the required dimensional accuracy or surface finish for a particular component feature and so should not be used.

Environmental sustainability

As a designer you should always try to specify the most sustainable materials and manufacturing processes that have the least environmental impact.

For example, powder coating or water-based paints should be specified in preference to traditional paints that contain solvents with high levels of volatile organic compounds (VOC). VOCs can be harmful to workers during application and cause unnecessary environmental pollution.

Design for manufacture and assembly

The majority of design engineers have to work within the constraints of their company's manufacturing and assembly facilities. It is important to bear this in mind when designing new products.

- New tooling and equipment is expensive and it takes time to specify, manufacture/purchase, install and train workers to use them safely.
- Wherever possible existing tooling, expertise, experience and manufacturing capability should be used in preference to specifying an entirely new or unfamiliar process.
- Similarly, designs should consider ease of assembly. For instance, placing fixings and fasteners in positions that are difficult to access increases assembly time and adds unnecessary cost.

Matching equipment and design features

A company has a CNC punch press used for piercing sheet metal blanks. The tooling head contains eight automatically interchangeable punches of different shapes and sizes. When designing new sheet metal products, features will be designed so that they can be manufactured using a combination of these eight standard punches. This will cut tooling costs and reduce set-up time.

Now try this

Using the surface finish process selection chart provided above, name the two processes capable of providing a surface finish roughness average (Ra) of 0.3 μm.

Generating design ideas

There are several approaches that you can take to generating ideas when you need to **modify** or **improve** an engineering product in line with **a customer's requirements**.

Technical design criteria

When you need to come up with ideas to improve an aspect(s) of a product, you will need to address particular design criteria established from the customer brief. Some common technical drivers are shown opposite.

Common technical drivers

Design improvement can be driven by considerations such as:

- preventing corrosion
- increasing wear resistance
- preventing recurrent failure
- keeping up with technological advancement
- material cost reduction
- manufacturing cost reduction
- performance improvement
- new features.

Generating ideas

Once you have established the key design criteria there are a wide range of idea-generation techniques that will help inspire you and encourage creativity. At this stage it's all about getting your ideas down on paper. Don't worry about the practicalities.

Techniques for generating ideas

You can generate ideas by brainstorming, mind-mapping, morphological analysis, SCAMPER and systematic search method.

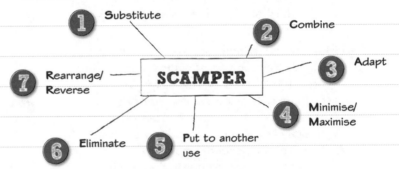

1. Substitute
2. Combine
3. Adapt
4. Minimise/Maximise
5. Put to another use
6. Eliminate
7. Rearrange/Reverse

SCAMPER

Initial design ideas

To narrow your thoughts down into initial design ideas you should consider the criteria below. The initial idea that answers these questions best can be taken forward to the next stage and developed further.

- Is the idea fit for purpose? Can it provide the improvement required?
- Can it be achieved within applicable constraints? Will it fit in the space available? Is it potentially dangerous?
- Can one or more initial thoughts be refined or combined to provide an improved design idea?
- Does it meet the client brief?

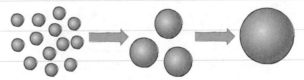

Idea generation   Initial ideas   Developed design proposal

Once you have generated a range of ideas (usually 10+), narrow them down to initial ideas considered in greater depth and presented in more detail.

Now try this

The owner of a canoe wants to be able to transport it using their family car. They do not have a tow bar and so cannot use a trailer. They don't want to have a permanently mounted roof rack on the vehicle or make any other permanent modifications.

Use one of the techniques above to generate five sketched ideas with annotations outlining different ways this might be achieved.

Development

Once you have identified the initial idea that best fits the desired outcomes of a product design specification (PDS) you need to take it forward and develop it in significantly more detail.

Design development

When developing your proposed design solution you are trying to bring every aspect in line with the requirements of the PDS. You should consider the following elements, which are a good guide to what should be included, although not exhaustive.

1 **Aesthetics.** How important is the aesthetic appeal of the product? Does it need to reflect a certain established brand image? Will it appeal to the product's target market?

2 **Ergonomics.** Can you improve how the product interacts with its intended users? Is it easy to handle, move, store and operate? Will a single size variant be sufficient or are a range of sizes needed to reflect user requirements?

3 **Sizes.** Does the product comply with given limitations on size? Are there established sizing conventions applicable to the product?

4 **Mechanical principles.** Are mechanical elements robust and of sufficent strength, stiffness or flexibility? What are the maintenance requirements? Are mechanisms able to operate the product with sufficient speed, force and control? Are there any potential mechanical hazards that need to be designed out or guarded against?

5 **Electrical and electronic principles.** Can the operation of the product be improved using electronic control and sensor technology? Is it an electrical power source, a fixed external mains supply or a portable internal system such as a battery? Can you minimise electrical power consumption to increase efficiency? Are there any potential electrical hazards that need to be designed out or guarded against? Do electrical systems need protection for the product's working environment?

6 **Materials.** Are the materials the most appropriate to provide the necessary physical, thermal, mechanical and electrical properties? Are they sustainable? Can recycled materials be used? Are the materials cost effective?

7 **Manufacturing processes.** How will the product be manufactured? Are there features of the design that prevent the use of certain manufacturing processes? Can the design be optimised to use less expensive and/or more sustainable manufacturing techniques?

8 **Assembly.** Can the number of components be reduced to simplify assembly? Can the use of clever design incorporate several features into a single component? How will components be fixed together?

9 **Cost.** Have you made the most efficient use of materials? Is investment in new tooling or equipment required?

10 **Safety.** Are there any hazards present that might cause harm during manufacture, use, recycling or disposal? Can these be designed out or easily mitigated?

11 **Sourcing components.** What kinds of fasteners or other applicable technologies are available from external suppliers? Have you made best use of off-the-shelf components? Will specialist components need to be sourced from external suppliers or manufactured in-house?

Now try this

Consider the ergonomic requirements of a mountain bike and outline three aspects of the design that you might investigate and review during a product improvement redesign.

It's essential to get the ergonomics of a mountain bike right if it's going to be comfortable and easy to ride.

Design information

During the design process it is essential to inform your choices by using **technical information** from high-quality, accredited sources.

Information sources

To inform your development work you'll need to refer to various sources of information. This will be especially useful when selecting materials and components.

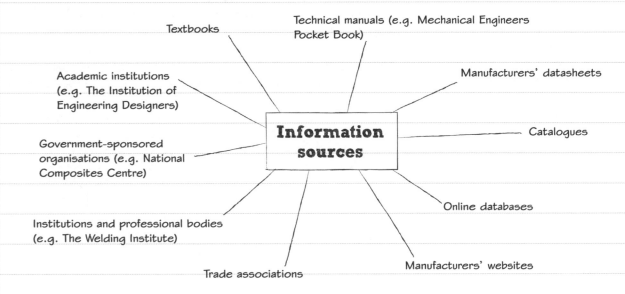

Textbooks

Technical manuals (e.g. Mechanical Engineers Pocket Book)

Academic institutions (e.g. The Institution of Engineering Designers)

Manufacturers' datasheets

Government-sponsored organisations (e.g. National Composites Centre)

Information sources

Catalogues

Institutions and professional bodies (e.g. The Welding Institute)

Online databases

Trade associations

Manufacturers' websites

A note of caution!

Beware of relying on open-source information published on the internet. Many popular and undoubtedly useful websites such as Wikipedia are open source. This means that anyone can contribute to the information on the site and, as a consequence, also means that there is no guarantee that any of it is wholly accurate.

It is good practice to find several corroborating (confirmed) sources for important or safety critical information. Ideally, these will be provided by accredited professional organisations who will have checked and verified their published material.

Don't rely too heavily on the internet. Technical reference books contain a wealth of information and are available in every library.

Now try this

Nichrome is a metal alloy containing 80% nickel and 20% chromium that is widely used in the manufacture of heating elements for electric ovens and toasters.

Suggest three reliable sources you might use to find the electrical resistivity of nichrome wire.

139

Freehand sketching, diagrams, technical drawings

Design concepts can be communicated quickly and easily using freehand sketching techniques. Technical information can be best conveyed using an appropriate diagram or technical drawing.

Freehand sketching 2D

This represents details of one face of a component and, as such, can convey only limited information about the component as a whole. You would use a quick 2D sketch to convey simple information quickly.

Freehand sketching 3D

You can use 3D techniques to provide a more realistic representation of a product or component. A great deal of visual information can be conveyed in a 3D sketch.

3D sketching techniques include isometric, oblique, cabinet oblique, single-point perspective and two-point perspective.

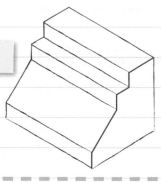

Example of isometric sketching.

Illustrations

An illustration is meant to provide a highly realistic impression of a **finished product** rather than convey technical information. You should use illustrations when the **form** of a product is of primary importance; for example, when designing a new car.

Diagrams

You can use diagrams to convey **schematic information** that helps to explain how a product works.

Diagrams can include:

- electrical circuit diagrams
- hydraulic/pneumatic circuits
- wiring diagrams
- flow charts.

Technical drawings

Technical drawings convey the information required to **manufacture** a component or product.

Technical drawings can include information on:

- dimensions
- tolerances
- surface finish
- materials
- explanatory notes.

Orthographic projections

A technical drawing is laid out in a **series of views** arranged as first or third angle orthographic projections. You will need to be familiar with how to read and draw orthographic projections.

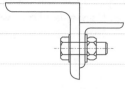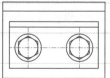

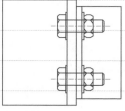

Third angle orthographic projections are most commonly used in the UK.

Now try this

Take the example of an isometric sketch given above and reproduce it as a third angle orthographic projection.

Graphical techniques

A range of graphical techniques can be used to communicate **technical** and **design information** to make it easier to understand and to access.

Keys are often used with charts, diagrams and other graphical techniques where symbols or colour coding have been used. The key explains the meaning of the symbols and colours used.

Charts are often used to display numerical data graphically, making it clear and easy to understand.

Shading – Contrasting colours on charts and diagrams make them stand out and easy to read.

Conventions – Technical drawing conventions, including the use of standard symbols, are laid down in national and international standards. These standards are, in part, a graphical dictionary setting out the agreed definitions for the symbols used. In the UK, BS8888 (2013) is the current standard that specifies how a technical product specification, including drawings and a range of other technical documents, should be completed.

Graphical techniques

Animation – Can add a little interest to a PowerPoint presentation but at its most powerful it can provide a virtual flythrough of a complex product in operation. Many 3D computer-aided design (CAD) software packages incorporate powerful animation capabilities that show potential customers how a product might look and operate long before anything is ever actually manufactured.

Symbols – Standard symbols have been developed for a wide range of indivdual components (electrical, electronic, pneumatic/ hydraulic) and engineering features (surface finishes, centre lines, threads). They enable engineers and designers across the globe to communicate in a common graphical language.

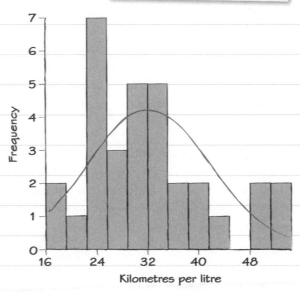

Histogram with normal curve

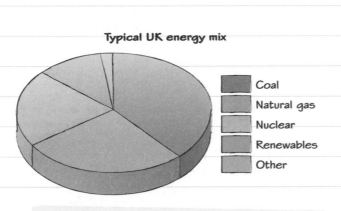

Typical UK energy mix

Key: Coal, Natural gas, Nuclear, Renewables, Other

Pie charts (showing the use of a key here) and histograms are common and familiar methods of graphically representing data.

The top five areas of UK national government expenditure have been estimated in the past as welfare (25.3 %), health (19.9 %), state pensions (12.8 %), education (12.5 %) and defence (5.4 %).

Use an appropriate graphical technique to illustrate the data so that it can be compared easily.

Written communication

As well as communicating through graphs, diagrams and charts, etc. you must also provide clear **written explanations** to explain and expand upon your ideas, thoughts and design decisions.

Written documentation

As a design engineer you will be expected to provide a range of **clear** and **accurate** written documents. Some types of engineering documentation are shown below.

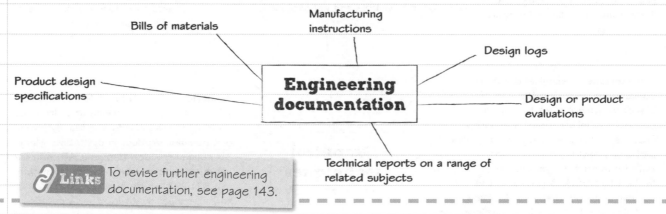

Bills of materials

Manufacturing instructions

Design logs

Product design specifications

Engineering documentation

Design or product evaluations

Technical reports on a range of related subjects

Links To revise further engineering documentation, see page 143.

Technical language

When writing annotations, notes and written explanations you should always use formal technical language.

Your technical language should:

• be clear

• be concise (using short sentences)

• be logical

• be well structured

• be factual

• be accurate

• support any visual elements (e.g. graphs, charts and diagrams)

• use appropriate specialised terms

• use accurate spelling and grammar.

Annotation

Any sketches, diagrams or illustrations you use should be accompanied by effective annotation.

Your annotation should:

• provide information

• be concise

• provide explanation

• provide context

• identify areas of importance

• form links between features and ideas.

Interpreting results

Often data collected from product testing, feedback from in-service performance or market research will influence your design decisions. **Interpretation** of the results and how they link with proposed design changes needs to be communicated clearly and systematically.

Your data interpretation should:

• be clear

• be justified

• be objective

• consider possible sources of error.

• reflect the data

Now try this

Rewrite the following paragraph relating to a 3-pin mains electrical plug, using appropriate technical language:

The plastic bit on the back of the plug smashed up into loads of bits when it got dropped onto the floor. The stuff it was made of was too easy to crack and wasn't bendy at all. It needs replacing with something softer that won't break.

Look at the information on material properties pages 105–107 and modes of failure page 111 if you need a reminder of the kind of language you need to use.

Design documentation

In order to fully specify a design solution you will need a range of written information, drawings and other diagrams. It is also good practice to keep a design log so a record is available of all the decisions, refinements and alterations made during development.

Design documentation

Design documentation will include some or all of the following:

Component detail drawings – Every manufactured component needs a detailed engineering drawing. This is usually a first or third angle orthographic drawing containing technical information such as dimensions, tolerances, finishes, materials etc.

General assembly drawings – Often an exploded isometric projection, these show how all the parts in an assembly fit together. Individaul parts are numbered and their part reference and description given in a corresponding table.

Material specifications – The material used for each manufactured component must be specified. This might include the material name, type or grade, manufacturer, size and form.

Component specifications – Other bought-in components must be specified. This might include component name, description, manufacturer's part number.

Bill of materials – All bought-in components, manufactured components and raw materials are included in a parts list known as a bill of materials.

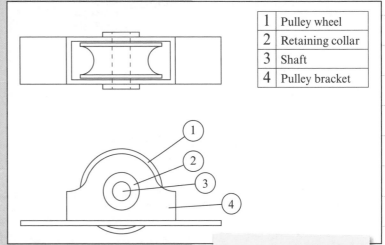

1	Pulley wheel
2	Retaining collar
3	Shaft
4	Pulley bracket

An example of a general assembly drawing.

Production plan – All the steps required to manufacture individual components, subassemblies and then the finished product need to be specified and explained.

Electronic circuit diagrams – Any printed circuit board (PCB) included in the product will require an associated circuit diagram.

Electrical wiring diagrams – Any electrical components wired together in the product will require an associated wiring diagram.

Flowcharts – Flowcharts are often a useful way of presenting sequential operations such as assembly or testing procedures.

Now try this

The design documentation for a new product contains errors that have not been noticed during the usual checking procedures.

What will be the consequences during production of:

1 an essential bought-in component being missed off the bill of materials
2 a hole being drawn smaller than the required size.

Iterative development

You will find the design process is not generally linear, where a series of consecutive steps lead from the product design specification (PDS) to a finished product. Instead, **several cycles** of development take place throughout a process of **continual refinement**.

Refinement

You can improve a product or process by analysing its performance in comparison to the requirements of the original PDS. This might happen several times during the design process.

Refining a product, process or design involves analysing it to find ways of improving it.

The iterative design cycle

The design cycle starts its life with a design brief from a client. Ultimately, through a cyclic process, it will lead to a finished product that meets all the client's requirements.

There is no limit to the number of iterations it might take to fully optimise a design solution. However, in practice short lead times and development cost limitations mean that compromises often have to be made in order to launch a product on time and to budget.

Now try this

An example of an iterative approach or a product that is constantly evolving by iteration is the mobile phone.

Suggest other examples of products where an ongoing iterative design process has made regular updates and product improvements possible.

Statistical data 1

Engineering decisions are often based on statistical data and you need to be familiar with different types of data and how measurements of central tendency (mean, median and mode) can provide useful insight into a problem.

Discrete variables

Discrete variables can take only certain **defined separate values**; e.g. the number of components in an assembly or the number of equipment breakdowns in one month or the number of workers on a production line.

128	128	129	125	128
128	129	128	128	129
127	98	127	127	129
128	96	129	115	128
129	110	127	120	

Table showing the hourly output of a production line (discrete variable).

Continuous variables

Continuous variables can occupy **any value within a certain range**; e.g. the time taken to manufacture a component, measured component dimensions or engine running temperature.

52.19	51.00	53.58	51.87	51.80
52.80	52.29	52.11	52.97	53.15
52.04	52.01	52.24	52.73	52.51
51.64	52.90	52.85	51.83	52.60
51.00	51.37	52.52	52.24	52.25

Table showing quality inspection measurements of machined shaft diameters (continuous variable).

Mean

The mean is the **numerical average** of a data set. It can be found using this formula:

$$\bar{x} = \frac{x_1 + x_2 + x_3 + \ldots + x_n}{n}$$

where:

x is the variable

$\bar{x}$ is the mean

n is the number of variables in the data set.

Median

The median is the **middle value** in a data set when arranged in ascending order of x. To calculate the median, use these formulae:

Median (when n is odd) $= x_{\frac{(n+1)}{2}}$

Median (when n is even) $\frac{x_{\frac{n}{2}} + x_{\frac{n}{2}+1}}{2}$

where:

x is the variable

n is the number of variables in the data set.

Worked example

An engineer measures a set of discrete variables of the breakdowns for seven months and finds the following results:

12	11
13	12
10	9
8	

Find the mean value of the data.

$$\bar{x} = \frac{(12 + 11 + 13 + 12 + 10 + 9 + 8)}{7}$$

$$\bar{x} = \frac{75}{7}$$

$$\bar{x} = 10.71$$

Mean = 10.71

This method is also used for continuous variables.

Worked example

Find the median for the following set of values:

8, 9, 10, 11, 12, 12, 13

In this case the number of variables is odd, so use:

Median $= x_{\frac{(n+1)}{2}}$

Median $= x_{\frac{(7+1)}{2}}$

Median $= x_{\frac{8}{2}} = x_4$

Median is x_4 or the 4th number, which is 11.

Now try this

Calculate the mean for the set of discrete data or the continuous data given above.

Statistical data 2

Other statistical measurements you need to be familiar with are **mode**, **variance** and **standard deviation**. Mode is another way of looking at central tendency whereas variance and standard deviation measure the spread of the data about the mean.

Mode

The mode is simply the data value that **occurs most frequently** in a data set. Mode is generally more useful when dealing with discrete data.

For example, in this data set 12 is the most common value as it appears twice:

8, 9, 10, 11, 12, 12, 13

So the mode = 12.

Standard deviation (σ)

Standard deviation is a measure of the **spread of data** and is closely related to **variance** (see below).

Standard deviation = $\sqrt{\text{Variance}}$

 Links To revise standard deviation in more detail, see page 149.

Variance

As well as central tendency, it is also useful to measure the **spread of data** about the **mean** to indicate the amount of **variation** in a data set.

For this you can use the variance formula:

$$\text{Variance} = \frac{(x_1 - \bar{x})^2 + (x_2 - \bar{x})^2 + \dots + (x_n - \bar{x})^2}{(n-1)}$$

where:

x is the variable

$\bar{x}$ is the mean

n is the number of variables in the data set.

Worked example

Calculate the variance for the data set below.

12	11
13	12
10	9
8	

Remember that you calculate the mean by adding up all the variables in the data set and dividing the answer by the number of variables.

Be careful with your calculator when calculating the square of a negative number. Remember $(-2)^2$ is different to $^22^2$ on most calculators.

$$\text{Variance} = \frac{(x_1 - \bar{x})^2 + (x_2 - \bar{x})^2 + \dots + (x_n - \bar{x})^2)}{n - 1}$$

$$\text{Variance} = \frac{(12-10.71)^2 + (11-10.71)^2 + (13-10.71)^2 + (12-10.71)^2 + (10-10.71)^2 + (9-10.71)^2 + (8-10.71)^2}{7 - 1}$$

$$\text{Variance} = \frac{(1.6641 + 0.0841 + 5.2441 + 1.6641 + 0.5041 + 2.9241 + 7.3441)}{6}$$

$$\text{Variance} = 3.238$$

Now try this

Calculate the **standard deviation** of the data in the example above.

Data handling and graphs 1

With any given or collected raw data it's often necessary to carry out some form of **processing** and/or **calculations** before it can be displayed graphically and interpreted.

Raw data

Raw data will not be in numerical order or have been gathered into groups of similar values or classes.

51.00	52.16	52.59	52.73	53.48	53.89
51.00	52.01	52.63	52.80	53.37	54.49
51.57	52.04	52.70	52.85	53.29	54.26
51.64	52.11	52.51	53.00	53.10	
51.80	52.19	52.52	53.32	53.62	
52.06	52.24	52.60	53.15	53.52	

Table of raw data measurements from a machined component.

Frequency tables

To make continuous data easier to interpret, classes representing a range of values are used. The number of data points falling into each class is counted and recorded as a frequency. The cumulative frequency is just the sum of consecutive frequencies.

Class	Frequency	Cumulative Frequency
51.00–51.49	2	2
51.50–51.99	3	5
52.00–52.49	7	12
52.50–52.99	9	21
53.00–53.49	7	28
53.50–53.99	3	31
54.00–54.49	2	33

Frequency and cumulative frequency table of measurements from a machined component.

Frequency histogram

You can plot **frequency** in each class as a histogram to help analyse how the data are distributed.

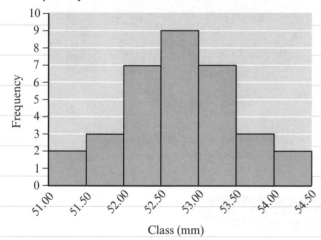

Class (mm)

Frequency distribution histogram of measurements taken from a machined component.

Cumulative frequency histogram

You can plot **cumulative frequency** as a histogram to help analyse how the data is distributed.

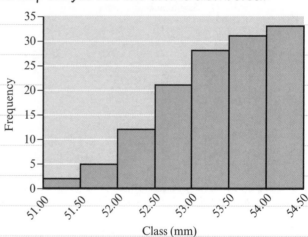

Class (mm)

Cumulative frequency distribution histogram of measurements taken from a machined component.

Now try this

Analyse the data provided below by compiling a frequency distribution table and plotting a frequency histogram.

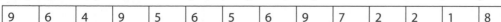

9	6	4	9	5	6	5	6	9	7	2	2	1	8

This is discrete data and so there is no need to split it into classes. Just count the number of instances of each value between 1 and 9.

Data handling and graphs 2

Ways of displaying data to make it easier to interpret are shown below, following on from those revised on page 147.

Bar charts

A large amount of discrete data displayed in a table is virtually meaningless until it has been processed and a frequency distribution table has been constructed.

Even then it can be difficult to get an overall impression of the distribution until it is plotted graphically.

For example, the frequency distribution in the table below becomes a lot easier to interpret when displayed as a bar chart.

Month	Workplace accidents
March	9
April	7
May	7
June	2
July	2
August	1

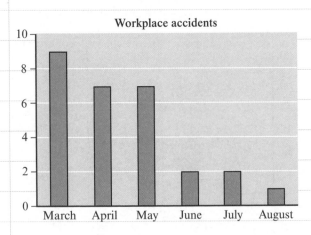

Workplace accidents

Bar charts are generally used to display discrete data. Unlike a histogram, a bar chart has spaces between the columns.

Pie charts

Even simple data displayed in a table lacks impact and it can be difficult to visualise how different values compare.

For example, the contents of the table below have significantly more impact when displayed as a pie chart.

Country	Manufacturing output (billions of $)
USA	2310
Japan	2100
China	1660
Germany	468
UK	312

Manufacturing output (billions $)

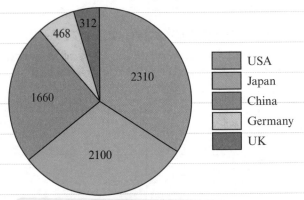

Pie charts are an effective way of illustrating comparisons.

Now try this

Use an appropriate chart to display the following discrete data on machine breakdowns over a 6-month period.

Month	Machine breakdowns
Jan	9
Feb	3
Mar	6
Apr	8
May	1
Jun	4

Frequency distributions

You need to be familiar with forms of frequency distributions that occur frequently in engineering problems. The most important of these is the **normal distribution**.

The normal curve

Often in engineering processes, data is spread evenly on both sides of the mean – when plotted graphically it shows a distinctive bell-shaped curve. This is characterised mathematically by what is known as the **normal distribution** and is encountered so often that you need to be aware of its implications.

The normal distribution is **symmetrical** about the mean, and mean, mode and median share the same value.

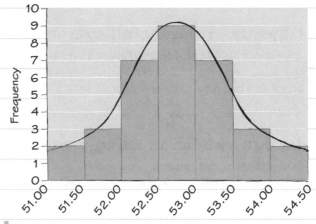

Standard deviation (σ)

The standard deviation of a normal distribution is the square root of its variance.

Standard deviation is useful because it provides a measure of the **extent of variation** that exists in a distribution. The lower the standard deviation, the tighter the normal curve and the lower the variation. In engineering, minimising variation is vital.

For example, a manufacturing process that is properly under control will have a standard deviation low enough to ensure that a batch of components will all (well 99.99% at least) be within the required dimensional tolerances.

Frequency distribution histogram of measurements from machined components showing the distinctive bell curve of the normal distribution.

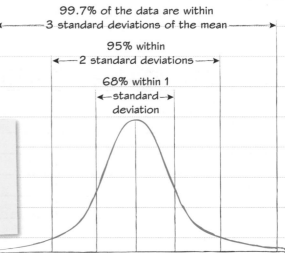

In the normal distribution 68% of all values lie within one standard deviation either side of the mean, 95% within two standard deviations either side and 99.7% within three standard deviations either side. For example, if the mean is 100mm and the standard deviation is 0.25 then 68% of all data will fall between 99.75 and 100.25mm.

Skewed distributions

Some distributions are **not symmetrical** and do not follow the normal model. In skewed distributions the mean value becomes less relevant and the median provides a better indication of central tendency.

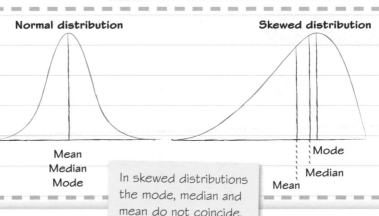

In skewed distributions the mode, median and mean do not coincide.

Now try this

A set of measurement data has a normal frequency distribution. The mean is 48.8 mm and the standard deviation is 0.17.

Find the upper- and lower-size limits between which 99.7% of the inspection measurements are likely to fall.

Validating the design

Once a design solution has been developed, it is important to validate it by **referencing** (comparing) its form, function, performance and other characteristics against the original requirements of the product design specification (PDS).

Objective referencing against the PDS

It is important that you are **honest** and **objective** when assessing how well your design solution meets the requirements of a given PDS. This is a vital part of the development of your design into a finished product that fully satisfies your client.

Design validation using a weighted matrix

In practice some criteria required by a PDS might negatively impact on others. For example, increasing battery life might mean compromising on overall product size and weight to accommodate a larger battery. Often, a **compromise solution** where one criterion is given priority over another has to be reached, and these difficult design decisions are made easier by analysing a design solution using a weighted matrix.

- A weighted matrix allocates a numerical weighting to each criteria of the PDS reflecting its relative importance.
- The sum of all the weightings allocated should come to 1.
- Each of the criteria is then given a score out of 10 to reflect how well the design solution reflects the requirements of the PDS.
- This score is then multiplied by the weighting to give a weighted score for that criteria.
- When the weighted scores for each criteria are added together then an objective measurement of how successfully the PDS has been met is expressed as a total score out of 10.

Evaluation criteria from PDS	Weighting	Score (out of 10)	Weighted score
Performance	0.3	8	2.4
Comfort	0.3	9	2.7
Aesthetics	0.1	10	1.0
Weight	0.2	8	1.6
Cost	0.1	6	0.6
		TOTAL (out of 10)	**8.3**

Now try this

1 Calculate the weighted matrix score for the design solution being validated in the example below.

Evaluation criteria from PDS	Weighting	Score (out of 10)	Weighted score
Performance	0.1	9	
Comfort	0.4	4	
Aesthetics	0.2	6	
Weight	0.2	9	
Cost	0.1	8	
		TOTAL (out of 10)	

2 Which of the five areas evaluated should be the main focus of further design efforts?

Benefits and opportunities

During validation you should be aware of how to manage the potential benefits that further development might bring, and weigh these against financial and other constraints.

Balancing benefits and opportunities against constraints

Not all the potential benefits of developing a proposed design solution may be worth the compromises necessary in other areas, or the expense needed to achieve them.

Cost–benefit analysis: Products need to make commercial sense and any potential benefits must be weighed against the cost of achieving them. In general, once a certain level of performance is achieved even small further improvements are too expensive.

Environmental benefits: It may not always be possible to meet all the environmental and sustainability requirements of a design. In some instances the use of recycled materials, although giving obvious environmental benefits, may not be possible given the required mechanical characteristics of the components.

Benefits and opportunities vs constraints

Health and safety risks: Product or performance improvements must not be allowed to compromise user safety and you must consider any knock-on effects.

Product life cycle considerations: Compromises or constraints at one stage of the product life cycle may be offset by benefits gained in another. Design and manufacturing complexity might be offset by the ease by which materials can be separated during recycling at end of life.

The costs of supplying more power to a motorbike engine may be considered acceptable but this cannot be done safely without also making expensive upgrades to the suspension and braking systems.

Indirect benefits and opportunities

By considering how a product will be used it is sometimes possible to identify **indirect benefits** that your client or their target market might not yet appreciate. This provides opportunities for widening the appeal of a product and potentially increasing sales.

More benefits

The **direct benefit** of using a vacuum cleaner is that it is quicker and more efficient at cleaning floors than a traditional dustpan and brush. An **indirect benefit** is that fine dust, pollen, pet hair and other potential allergens are removed and physically isolated inside the vacuum. Reducing airborne allergens can help to lessen symptoms of hayfever or asthma.

Now try this

The performance of a desktop computer you are designing would be improved by installing a new type of central processor unit (CPU) onto the motherboard. However, the alternative CPU requires more power and generates more heat than the existing chip and has a different pin configuration.

Which three elements of your system design would you need to investigate as part of a feasibility and costing study into this potential change?

Further modifications

During validation further modifications may be made to take into account proposed manufacturing methods and other technological developments.

Design for manufacturing

Even at a late stage in the process there will be scope for improving the design to help make it more suitable for the manufacturing processes to be used.

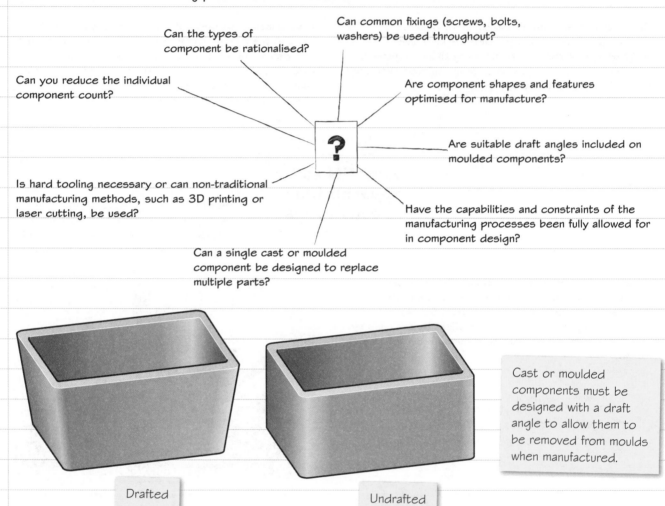

Can the types of component be rationalised?

Can common fixings (screws, bolts, washers) be used throughout?

Can you reduce the individual component count?

Are component shapes and features optimised for manufacture?

Are suitable draft angles included on moulded components?

Is hard tooling necessary or can non-traditional manufacturing methods, such as 3D printing or laser cutting, be used?

Have the capabilities and constraints of the manufacturing processes been fully allowed for in component design?

Can a single cast or moulded component be designed to replace multiple parts?

Drafted

Undrafted

Cast or moulded components must be designed with a draft angle to allow them to be removed from moulds when manufactured.

Technology-led adaptations

During the development process, **emerging** manufacturing or processing technologies could become available that might apply to the product being designed. Alternatively, **changes** in related technologies that the product being designed would be expected to use or interface with might lead to **necessary design adaptations** to insure against premature obsolescence.

Design adaptations – planning for obsolescence

Manufacturers of televisions for the UK began to incorporate dual digital and analogue receivers long before the planned switch-off of the analogue signal. This reassured customers that their new TV would still work when the switch over to digital was made.

Now try this

Describe a potential problem that the inclusion of a draft angle on cast metal components might cause during subsequent machining operations.

Your Unit 3 set task

Unit 3 will be assessed through a task, which will be set by Pearson. You will complete a task to demonstrate your knowledge and understanding of engineering product design and manufacture as you follow a standard development process of interpreting a brief, scoping initial design ideas, preparing a design proposal and evaluating the proposal.

Revising your skills

Your assessed task could cover any of the essential content in the unit. You can revise the unit content in this Revision Guide. This skills section is designed to **revise skills** that might be needed in your assessed task. The section uses selected content and outcomes to provide an example of ways of applying your skills.

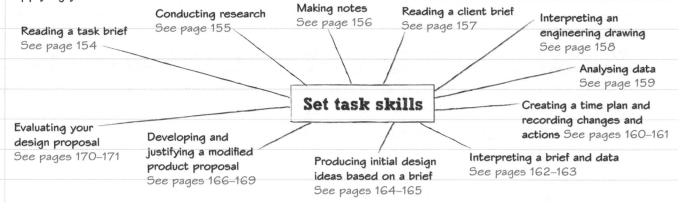

Reading a task brief
See page 154

Conducting research
See page 155

Making notes
See page 156

Reading a client brief
See page 157

Interpreting an engineering drawing
See page 158

Analysing data
See page 159

Set task skills

Creating a time plan and recording changes and actions See pages 160–161

Interpreting a brief and data
See pages 162–163

Producing initial design ideas based on a brief
See pages 164–165

Developing and justifying a modified product proposal
See pages 166–169

Evaluating your design proposal
See pages 170–171

Workflow

The process of the task might involve you in the following steps:

☑ Read a task brief, conduct research and make notes.

☑ Read further information and interpret a client brief, engineering drawing and data.

☑ Create a time plan and record changes and actions.

☑ Interpret a brief's operational requirements.

☑ Produce a range of initial design ideas based on the brief.

☑ Develop a modified product proposal with relevant design documentation.

☑ Evaluate your design proposal.

Check the Pearson website

This section is designed to demonstrate the skills that might be needed in your assessed task. The details of the actual assessed task may change, so always make sure you are up to date. Check the Pearson website for the most up-to-date **Sample Assessment Material** to get an idea of the structure of your assessed task and what this requires of you.

Now try this

Visit the Pearson website and find the page containing the course materials for BTEC National Engineering. Look at the latest Unit 3 Sample Assessment Material for an indication of:

* the structure of your set task, and whether it is divided into parts
* how much time you are allowed for the task, or different parts of the task
* what briefing or stimulus material might be provided to you
* any notes you might have to make and whether you are allowed to take selected notes into your supervised assessment
* what you might need to take into the assessment; e.g. a black pen and a spare, HB or B pencil, ruler, eraser, drawing instruments and calculator
* the activities you are required to complete and how to format your responses.

Reading a brief

Here are some examples of skills involved when reading a task brief and task information.

Task brief and information

The task brief and information are used as examples to show the skills you need. The content of a task will be different each year and the format may be different.

When reading a task brief and information:

- read it several times
- underline key information
- plan the focus of your research.

Task brief

A manufacturer has been approached by one of its clients for whom it manufactures brackets used in tunnels. The client has asked the manufacturer to <u>optimise the design</u> of the product, which is a bracket.

You will research the <u>design</u> and <u>manufacturing requirements</u> that are relevant to the bracket and its application. Your research should consider:

- <u>existing designs</u> for brackets
- the <u>manufacturing processes</u> and <u>technologies</u> that are being used and possible <u>alternatives</u>
- the <u>health and safety</u> requirements for the manufacturing processes and technologies
- <u>environmental considerations</u>, including <u>sustainability</u>
- <u>material requirements</u> and suitable material <u>properties</u>
- any other <u>relevant factors</u>, such as ease of fitting.

> It is a good idea to underline key information in the task brief and task information you are given.

You will be given further information on the specific issues with the existing bracket that will allow you to <u>redesign</u> the bracket and <u>evaluate</u> your solutions against the issues.

Task information

The bracket is attached mechanically to a metal support within a tunnel. The <u>purpose</u> of the machined bracket is to provide <u>support</u> for a <u>high-voltage insulated electric cable</u>, which runs the <u>length of the tunnel</u>. The electrical cable goes through the <u>65 mm diameter hole</u>. The number of brackets that are needed depends on the length of the tunnel but they are <u>generally manufactured in thousands</u>.

Currently, the bracket is machined from <u>low carbon steel</u>, which has <u>a paint finish</u> applied to it.

Existing Cable Support Bracket

Length = 318 mm,
Width = 200 mm,
Height = 115 mm

> You could include research into other existing designs for brackets that support cables in tunnels, for example.

Now try this

List a range of sources of information that might be used to support research activities for this task.

Conducting research

Here are some examples of skills involved when carrying out research.

Planning my research - factors

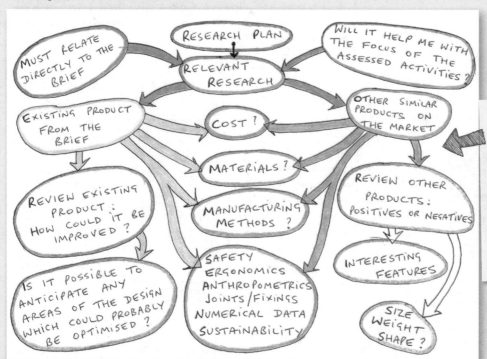

This diagram groups some relevant and useful factors when conducting **research**. You must decide the most important areas you need to know about based on your understanding of the task. Make sure you split your research task into all the **important areas**.

Steel castings:

- used when cast iron cannot deliver enough strength or shock resistance
- more difficult to cast than iron but has better wear resistance
- can withstand significant stress and strain without fracturing and can be further enhanced by alloying
- are much more difficult to machine than cast iron, and steel exhibits a lower damping ability, which can sometimes lead to excess vibration and noise (ringing or squealing), in certain applications.

Here is an example of some research into **manufacturing methods** in relation to a task brief relating to stable door hinges.

You need to identify **important facts** and could collate **data**, listing materials and their relevant properties.

Gather **relevant information** in your research.

Now try this

Make a note of key points it would be useful to research in relation to manufacturing methods for the task brief on page 154.

Making notes

Here are some examples of skills involved when making research notes.

Sample notes extract

The 'Ardcase pedal lock

Function
The device opens like a 'clam shell'. It is placed over the clutch, brake and accelerator pedals of the vehicle. It is then closed and locked using a 7 lever gun cabinet lock. This renders the vehicle undriveable.

Manufacture
The box is laser cut as a 'net' before being folded on a traditional folding machine in batches of 50. It is MIG welded along the seams. Outside surfaces are dressed with a hand grinder to smooth the corners. The product is shot blasted and powder coated. Internal pieces are held in place using a jig whilst being welded.

Materials
The whole casing is 4mm mild steel plate. Internal lock plates are 6mm steel plate which is hardened to prevent penetration using a drill. Lock is affixed using 4 x M5 socket head cap screws.

Cost
The retail price is £139 with a making cost of around £80

Health & Safety
Corners rounded to prevent injury but it is possible to trap fingers when closing the shell. Non toxic finish on exterior.

Sustainability: Mild steel is relatively easy to recycle. The product is designed to last a long time. Can be manufactured from re-used material.

Advantages/Disadvantages
Excellent deterrent against opportunist thief. Simple, effective and reliable. Needs secure storage inside the vehicle when not being used.

Data Total weight 4.5kg. Average fitment time 6 seconds Average removal time 5 seconds (with key!)

Finish
Oven baked powder dip coated

Aesthetics Bright colour as a visual deterrent. Aesthetics are second to function with this product

This extract shows some product analysis and research notes that respond to a brief relating to a security device to prevent theft of a vehicle.

Using subtitles is a useful way to ensure all the basic important points are covered. Each of the areas could be looked at in more detail on another research sheet if necessary. There may also be other information needed that is very specific to the topic being researched.

The subtitles can also act as a checklist of areas to be investigated, providing a good way to structure research of existing products:
- Function of the product
- Manufacturing methods and processes
- Costs
- Materials and material properties
- Health and Safety issues
- Sustainability and Environmental factors
- Relevant numerical data
- Aesthetics, Ergonomics, Anthropometrics
- Applied finish
- Advantages or Disadvantages of the design.

Now try this

Select a design for a can crusher (limit your selection to less than £10). Produce a research sheet, outlining some of the important, relevant information about your chosen product. Use the example notes above as a guide.

There are many products available to crush drinks cans, like the one shown here.

Reading further information

Here are some examples of skills involved when reading further task information and a client brief.

Further task information

When reading further task information, consider:

- how it continues from an earlier brief (see page 154), with a brief from a **client**
- additional information given in an **engineering drawing**; for example, see page 158 for an orthographic projection of the existing support bracket and a sectional view showing the bracket in operation (the position of the hole in relation to the wall and a metal support)
- additional information given in **numerical data**; for example, see page 159 for the numerical data that is referred to in the client brief.

Client brief

The client is aware that the current design has <u>a number of</u> issues, but the redesign has been triggered by the <u>bracket fracturing in service</u>. The client had intended the life cycle of the bracket to be 25 years. The client needs the manufacturer to identify the stage in the life cycle when the brackets begin to exhibit signs of fracture and design, and a <u>solution that will reduce the likelihood of the brackets fracturing in service</u>.

Based on simulations and testing, the client has provided the following information in Table 1, which can be used to perform a <u>statistical analysis of the service conditions</u> that the brackets are used in.

The client has asked the manufacturer to come up with an alternative solution that can also take into account the most <u>efficient use of materials and manufacturing processes</u>; however, the manufacturer also has an opportunity <u>to optimise the design in terms of form, sustainability and other factors</u>.

<u>The bracket must:</u>

- have a 65 mm hole with a dimensional tolerance of ±2 mm
- be manufactured so that the 65 mm hole is in the same position as the original bracket (see sectional view of the bracket in operation)
- have at least three mounting points.

A client brief outlines a client's expectations and requirements for a product. Read the client brief carefully and make sure you understand it. Here, the learner has underlined some key information.

Links Table 1 is shown on page 159.

Notice all the additional information you are given in the client brief. This will help inform your **redesign** of the product.

Links An **orthographic projection** and sectional view are shown on page 158.

Now try this

Look again at the client brief above, the orthographic drawing on page 158 and the data table on page 159.

1 What has triggered the redesign of the bracket?
2 What is the average life expectancy of bracket A?
3 What is the diameter of the six bracket mounting holes?

Use the orthographic projection drawing, the data table, and the task information all together, to find the facts you need.

Interpreting an engineering drawing

Here are some examples of skills involved when interpreting an **engineering drawing**, such as the one below, referred to in the client brief on page 157.

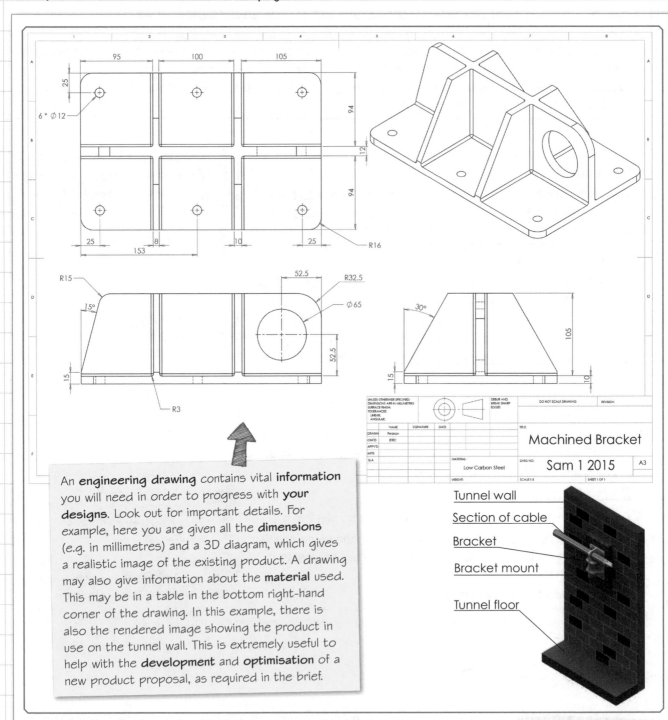

An **engineering drawing** contains vital **information** you will need in order to progress with **your designs**. Look out for important details. For example, here you are given all the **dimensions** (e.g. in millimetres) and a 3D diagram, which gives a realistic image of the existing product. A drawing may also give information about the **material** used. This may be in a table in the bottom right-hand corner of the drawing. In this example, there is also the rendered image showing the product in use on the tunnel wall. This is extremely useful to help with the **development** and **optimisation** of a new product proposal, as required in the brief.

Tunnel wall
Section of cable
Bracket
Bracket mount
Tunnel floor

Now try this

You will notice that none of this information is given directly on the drawing, but you can easily find out these essential measurements through simple calculation.

Using the drawing above, answer the following:

1 What is the overall length of the existing bracket?
2 What is the overall width of the existing bracket?
3 What is the overall height of the existing bracket?
4 What is the diameter of the cable hole?

Analysing data

Here are some examples of skills involved when analysing **data**, such as **Table 1** below, referred to in the client brief on page 157.

Bracket	Minimum temperature inside tunnel (degrees Celsius)	Maximum temperature inside tunnel (degrees Celsius)	Humidity inside the tunnel (percentage)	Distance from the tunnel entrance/ exit (metres)	Life cycle (years)									
					Test 1	Test 2	Test 3	Test 4	Test 5	Test 6	Test 7	Test 8	Test 9	Test 10
A	−10	18	40	20	25	27	15	22	30	24	26	19	24	28
B	−6	19	50	40	20	22	26	17	23	13	18	19	28	20
C	−2	22	60	60	15	17	11	13	16	14	21	9	27	16
D	0	26	80	80	7	19	16	11	7	1	9	14	23	11
E	3	30	80	100	8	4	7	16	12	22	6	7	2	9

Table 1 – Outcome of simulations and testing on the existing brackets

---

Initial analysis

- Bracket A and B appear to have the best life expectancy.
- They are closest to the tunnel exit or entrance.
- Humidity increases further away from the entrance/exit.
- The temperature increases further away from the entrance and exits.
- The combination of heat and humidity would appear to shorten the life expectancy of the brackets.
- The brackets appear to last longer in cooler, less humid areas of the tunnel.

In this example, the learner **analyses** the information in the table and **makes decisions** about which data is relevant to the brackets fracturing in service.

Some data can be understood 'at a glance' but some **calculations** may be needed to get the full picture. For example, here, the **average** life expectancy of each bracket will need to be calculated. You can do this by adding up the results of the tests for each bracket and dividing by the number of tests.

Now try this

Calculate the actual average life expectancy for each of the brackets.

Verify your 'at a glance' assumptions with calculation, to be sure you are correct.

Creating a time plan

When asked to create a short project time plan to carry out an iterative development process, consider each stage of the task and what it requires.

Sample response extract

Project time plan for the task

Produce the time plan for the full project, to carry out an iterative development process.

During the development process, record in my task booklet:

- why changes were made to the design during each session
- action points for the next session.

▼

Interpret the design brief into operational requirements, to include:

- product requirements
- opportunities and restraints
- interpretation of numerical data
- key health and safety, regulatory and sustainability factors.

▼

Produce a range of (e.g. three or four) initial design ideas based on the client brief, to include:

- sketches
- annotations.

▼

Develop a modified product proposal with relevant design documentation. The proposal must consider:

- a solution
- existing products
- materials
- manufacturing processes
- sustainability
- safety
- other relevant factors.

▼

Finally, evaluate:

- success and limitations of the completed solutions
- indirect benefits and opportunities
- constraints
- opportunities for technology-led modifications.

Demonstrate a **logical** and **iterative** approach to the design process. Action points should show forward planning clearly linked to the specifics of the product being redesigned, with consideration of what has happened in the previous session.

Make sure you meet the brief and consider enhanced product performance. The interpretation should include comments that extend the client brief into a set of cohesive operational requirements, not just a repeat of the client brief. Show your calculations and conclusions relating to the numerical data, commenting on them and taking action forward. Be specific about sustainability factors and health and safety, relating comments to the redesign of the product.

Ensure your ideas are feasible and fit for purpose, and reasonably different to the existing product and each other, when considering form and approach. Your adaptations should be major improvements when compared to the existing product and in relation to the brief. Be specific in your annotations, relating comments to the brief, so comments about, for example, materials or shapes, are applied in context rather than general.

Your response breaks down into two sub-tasks: a **solution** and **design documentation**.

- In relation to the existing product, the solution chosen and drawn should: be a clear improvement on the existing product (most importantly); show a clear variation in form/approach; be much safer to use and interact with.
- The **annotation and notes** should: refer to other products you researched and show how the features of different products have been used in the chosen solution; give reasons for the materials you selected for your solution; refer to your choices for sustainability and e.g. raw materials extraction, material and parts production, assembly, use and disposal/recycling.
- Make sure you use accurate terminology, and that any **documentation** allows a competent third party to manufacture the solution – e.g. include a reasonably accurate orthographic projection.

Give a balanced and thorough appraisal, with a sound rationale. Make sure you give good reasons as to why your solution is effective, with reference to the points in the brief. Include a focus on the opportunities, limitations and constraints of your chosen design solution. Be specific about further technology-led modifications, so your comments are applied in the context of your solution, rather than generally.

Now try this

Plan the time you need for each of the stages, using the information on timing and marks from the latest Unit 3 Sample Assessment Material on the Pearson website to guide you.

Recording changes and action points

Here are some examples of skills involved when recording the **changes** you make along with **action points**, during the **iterative development process**. The examples show responses to a brief relating to modifying a right angle bracket for outdoor use.

Carrying out an iterative development process

A project log is a document that records the progress made, key activities and decisions taken during the development of a project. Make sure that throughout the design response you:

- **justify changes** made throughout the development process to meet the requirements of the brief
- identify well-defined, logical and prioritised next **points of action**, that are clearly linked to the specifics of the product that is being redesigned, with consideration of what has happened in the previous session.

Sample response extract

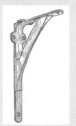 

Bracket could be manufactured from stainless steel.

In this example, a **sketch clearly records** the changes made to a design of a simple right-angle bracket manufactured from mild steel. The bracket is corroding after a few months of outdoor use and, as part of the development process, the drawing records the changes made to use stainless steel instead.

Sample response extract

Although stainless steel is more difficult to work with than mild steel, it may be a better material for the bracket because of its resistance to corrosion. The use of stainless steel would mean a corrosion-resistant finish would not need to be applied, unlike mild steel, which would need protection such as dip coating, electroplating or galvanising. Stainless steel has a naturally aesthetically pleasing finish, too.

This extract shows a clear understanding and use of the **iterative design process**. You need to **justify the changes** you make to improve your design during the development process.

This extract makes it clear how the proposed design change is **linked to research** and **the requirements of the design brief.**

Sample response extract

Link to research and brief

Manufacturing the bracket in stainless steel would fulfil the requirement of the design brief in terms of improving the life span of the product. The expected increase in cost would then be justified and may be acceptable to the client. My initial research indicated that many commercial products designed for constant use in harsh outdoor environments are manufactured from stainless steel (e.g. pipes on oil rigs).

Sample response extract

The action points identified in this extract are clear and logical.

Action points

I now need to take into account:

- the cost of stainless steel in relation to mild steel
- which grade of stainless steel would be most suitable
- whether different tools are needed to manufacture the bracket in stainless steel
- the expected life span of various stainless steels in a typical outdoor environment.

Now try this

Look back to the orthographic drawing of the cable bracket on page 158. A learner has decided to propose that the bracket should be manufactured in carbon fibre.

1 **Justify** why this may (or may not) be an improvement.

2 Give a list of **actions** that would be needed, to assess whether the proposal is realistic.

Interpreting a brief

Here are some examples of skills involved when interpreting a client brief.

For example, when interpreting a brief into **operational requirements**, include:

- product requirements
- opportunities and constraints
- interpretation of numerical data
- key health and safety, regulatory and sustainability factors.

Interpreting a brief into operational requirements

Consider how you will:

- interpret the brief into a cohesive and comprehensive set of **product requirements**, feasible **opportunities** and **constraints** that meets the brief and considers enhanced product performance
- show accurate calculation and interpretation of **numerical data**
- address key **health and safety, regulatory** and **sustainability** factors with relevance to the given context.

Worked example

The manufacturer of a car key remote wishes to upgrade the existing design. The new product must not cost more than the existing product, which is £25.

This short extract from a brief requires an upgrade to the existing design of a car key. You need to show you understand what the brief is asking you to do by **analysing** the different operational requirements. With your actual longer set task brief you may also need to interpret an **engineering drawing** and some **numerical data**, and apply any necessary calculations (see pages 154 and 157–159, for example).

🔗 **Links** To revise interpreting engineering drawings, see pages 140 and 158, and to revise interpreting numerical data, see pages 145–146, 159 and 163.

Sample response extract

<u>Regulations</u>	<u>Sustainability</u>	<u>Requirements</u>	<u>Constraints</u>	<u>Health and safety</u>	<u>Opportunity for improvements</u>
Must comply with legislation relating to radio transmission power and frequency...	Must be manufactured with at least 80% recyclable materials and components...	Must transmit a distance of at least 50m from the vehicle...	Must not cost more than the existing product (£25)	Must have no sharp edges...	Must incorporate recent advancements in radio transmission security...

In these extracts that start to interpret the brief, important points have been identified for each of the six operational requirements. You will be expected to identify multiple points in each of the six categories and take these forward, along with conclusions from any numerical data and engineering drawings.

Now try this

You have been asked to design a new dispenser for a standard 24 mm x 66 mm roll of sticky tape, to use on a desk, in a busy office. The dispenser must have a retail price of less than £10 and be no bigger than 150 × 100 × 100 mm. Under each of the six categories in the sample response extract above, write down at least two important points relevant to the brief, that you would take forward.

Interpreting numerical data

Here are some examples of skills involved when interpreting numerical data. Show that you can understand and analyse the information, carrying out calculations if needed. Calculations should be accurate and conclusions should be commented upon/taken forward.

The table below shows data collated by a motorcycle racing team, who have been experiencing a number of chain failures on their racing motorcycles.

Table 1 – Outcomes from testing of existing motorcycle chains

Chain	Motorcycle engine size (cc)	Motorcycle engine torque (Nm)	Number of standing starts	Number of race miles completed	Life cycle of chain (racing hours)								
					Test 1	Test 2	Test 3	Test 4	Test 5	Test 6	Test 7	Test 8	Test 9
A	1000	120	80	900	19	20	20	20	20	21	18	21	19
B	800	100	190	754	9	8	9	10	6	9	7	7	9
C	600	79	110	1120	15	14	13	15	13	12	16	14	15
D	250	34	58	1199	31	40	36	31	29	31	28	37	32
E	125	21	175	850	11	11	11	12	9	10	12	13	8

This extract from a brief shows a table of chain failures. The owner of the motorcycle team wants to know why chains are failing and what is the main cause of chain failure. You need to look at the figures in the table and decide which data is relevant and which is not. You must make decisions based upon the **data supplied** and **not make assumptions**. For example, you may think that chain failure on the motorcycles is likely to be more of a problem on bigger, more powerful bikes, but looking at the data above, this is clearly not the case.

In looking at the data for chain A on the 1000cc motorcycle, it did 80 starts and covered 900 race miles.

Average distance of each race = 900/80 = 11.25 miles

When repeating this calculation for each chain, it is clear that chain B only did races of an average distance of 3.96 miles (the shortest average race distance of any chain), but still had the shortest life expectancy. It is therefore not necessarily race distance causing chain failure. In looking at the data for chain B, the chain was on the 800cc bike, which completed 754 race miles. This was the shortest race mileage of all the chains tested, but the life cycle in racing hours is the shortest. However, this chain was subjected to more standing starts than all the others (190 starts). It would be reasonable to deduce that it is the number of standing starts contributing to chain failure.

In this extract from a response the data has been **analysed** and the **main factor** contributing to chain failure has been **identified**. The conclusions would inform ways forward.

Now try this

When analysing the data in the table above, it would be useful to know the **average** life expectancy for each chain. This figure is not given in the data table. Calculate the average life expectancy for each chain.

For each chain, you need to calculate the sum of chain life cycle hours divided by the number of tests.

Producing design ideas 1

Here are some examples of skills involved when producing **initial design ideas** to modify a product in response to a client brief. For example, the range of initial design ideas you produce should be based on the client brief and include **sketches** and **annotations**.

Initial design ideas

Consider how you will:

- produce a range (e.g. three or four) of appropriate initial design ideas that address the **brief** and are reasonably different to the existing product shown and each other, when considering form and approach.
- communicate the ideas **concisely** and with **clarity**, with appropriate use of **technical terms** that link to the **brief**.
- ensure your ideas are **feasible** and **fit for purpose**, showing adaptations that are major improvements when compared to the existing product and the points in the brief.

Sample response extract

The annotated sketch below is an extract of one idea in response to a brief to modify a car wheel-brace so it can be made small for easy storage in a car boot.

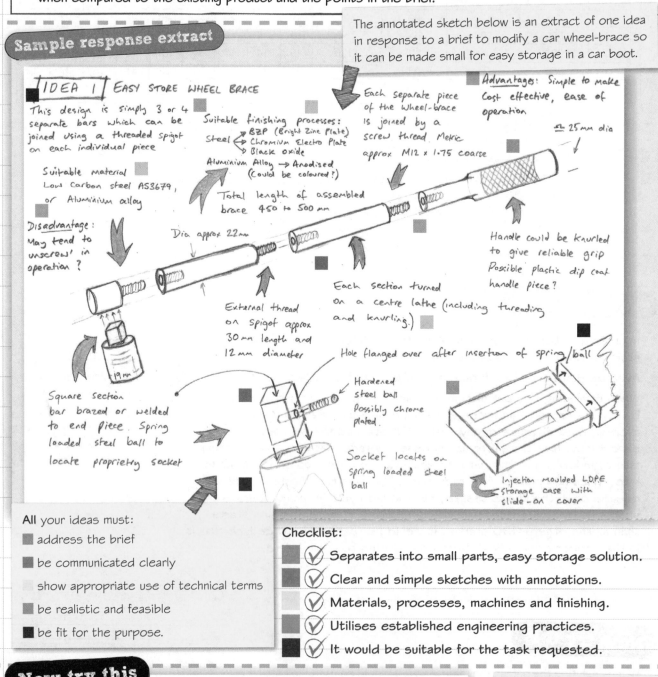

IDEA 1 EASY STORE WHEEL BRACE

This design is simply 3 or 4 separate bars which can be joined using a threaded spigot on each individual piece

Suitable material Low Carbon steel AS3679, or Aluminium alloy

Disadvantage: May tend to 'unscrew' in operation?

Suitable finishing processes:
Steel ⟨ BZP (Bright Zinc Plate) / Chromium Electro Plate / Black oxide
Aluminium Alloy → Anodised (could be coloured?)

Total length of assembled brace 450 to 500 mm

Dia approx 22mm

External thread on spigot approx 30 mm length and 12 mm diameter

Square section bar brazed or welded to end piece. Spring loaded steel ball to locate proprietry socket

19 mm

Each separate piece of the wheel-brace is joined by a screw thread. Metric approx M12 × 1·75 coarse

Each section turned on a centre lathe (including threading and knurling.)

Hole flanged over after insertion of spring/ball

Hardened steel ball Possibly chrome plated.

Socket locates on spring loaded steel ball

Advantages: Simple to make Cost effective, ease of operation

25mm dia

Handle could be knurled to give reliable grip Possible plastic dip coat handle piece?

Injection moulded L.D.P.E. storage case with slide-on cover

All your ideas must:
- ☐ address the brief
- ☐ be communicated clearly
- ☐ show appropriate use of technical terms
- ☐ be realistic and feasible
- ☐ be fit for the purpose.

Checklist:
- ☑ Separates into small parts, easy storage solution.
- ☑ Clear and simple sketches with annotations.
- ☑ Materials, processes, machines and finishing.
- ☑ Utilises established engineering practices.
- ☑ It would be suitable for the task requested.

Now try this

Sketch one idea, with annotations, for an adjustable bracket to hold a flat-screen computer monitor onto a wall.

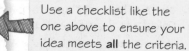

Use a checklist like the one above to ensure your idea meets **all** the criteria.

Producing design ideas 2

Clear sketches of your initial design ideas are important, although it is not your art skills that are being tested. Sketches must be understandable although every detail does not need to be shown at this stage. The notes and annotations with your sketches may provide the best method of demonstrating the important aspects of your idea and also your engineering skill and knowledge.

Sample response extract

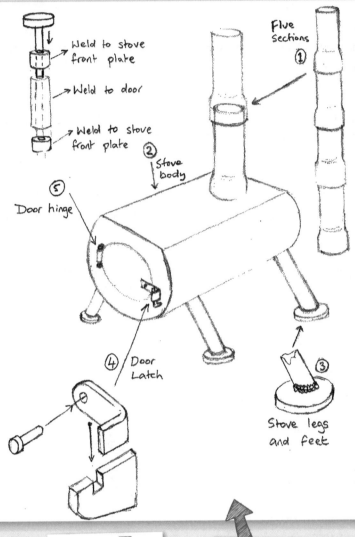

Flue Sections ①

Weld to stove front plate

Weld to door

Weld to stove front plate ②

⑤ Door hinge

Stove body

④ Door Latch

③ Stove legs and feet

① Cold drawn mild steel tube. 60 mm dia. Each section swaged to slide over next piece. All store inside stove body for transportation.

② Body fabricated from 4 mm low carbon steel plate. Bottom section rolled to shape and MIG welded to flat top cooking surface. Finished using high temperature, rust-inhibiting paint. Front panel with laser-cut door aperture MIG welded to front of body.

③ Legs from 25 mm seamless tubing. Welded to stove body and 4 mm flat mild steel disc at bottom. Tube cut to correct angle using power hacksaw.

④ Simple fabricated latch pivoted on steel pin (turned on centre lathe). Latch and receptor from 3 mm steel plate. Hole in steel plate door for pin, which is welded on inside surface.

⑤ 15 mm dia steel bar, bored 8 mm on lathe to suit 8 mm external dia steel pin turned on lathe. Bar parted into three pieces. Pin is case hardened after machining. Bar welded to door and stove body as indicated.

The sketch shows an extract from an initial idea for a camping stove. It would result in the finished product opposite, in response to a brief to modify a design for a camper's portable wood burning stove so it can be safely used for cooking and keeping warm. The extracts from annotations show the kinds of ideas that might be produced to accompany a sketch.

Now try this

You have been asked to design a smartphone holder for a car. Produce an initial sketch, with several annotations to accompany it.

> You could take into account some health and safety considerations.

Modified product proposal

Here are some examples of skills involved when developing a chosen idea (or combination of ideas) into a modified product proposal. This is an important part of the development cycle that draws on all aspects of engineering capability. For example, when **developing a modified product proposal with relevant design documentation**, the proposal must consider: a solution, existing products, materials, manufacturing processes, sustainability, safety, other relevant factors.

Develop a modified product proposal

Consider how you will break down your work to two sub-tasks: the **solution** and the **design documentation**. For example:

- **optimise** your solution, so it is a clear improvement on the existing product (the most important factor), demonstrating a **justified variation** in form and/or approach from the brief, and is much safer to use/interact with than the existing product.
- ensure your design proposal is **informed**, based on a thorough understanding of **existing alternative products.**
- ensure your selection of **material(s)** and **manufacturing processes** are **appropriate** to the brief and **justified** by balanced investigation of options, giving reasons for selection.
- take account of **sustainability** at all stages of the product life cycle.
- clearly reference the **safety** of the design.
- produce a range of **relevant formal documentation** to communicate the solution effectively.
- **annotate** the solution to be **concise**, to allow a competent third party to interpret effectively how to manufacture the solution.
- use **technical terminology** throughout.

Sample response extract

1 The example starts with the basic original **proposal** for the stove door latch, listing the **advantages** and **disadvantages** of the design.

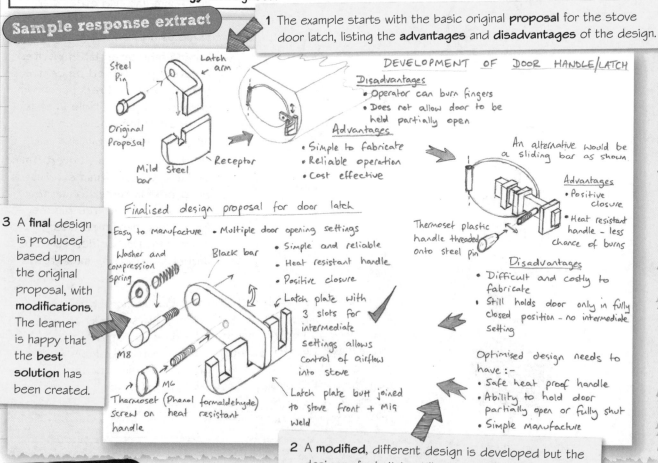

3 A **final** design is produced based upon the original proposal, with **modifications**. The learner is happy that the **best solution** has been created.

2 A **modified**, different design is developed but the designer feels it is still not the optimised solution.

Now try this

Look at the design for the stove legs on page 165. They are welded solidly to the stove body. Sketch a proposal that would allow the legs to either detach or fold up to the stove body, for ease of transportation. List the advantages and disadvantages for your proposal.

Justifying a design

As you develop a product to satisfy a client brief, you will make decisions relating to **material** choice, method of **manufacture** and the actual **design** of every element of a proposal. You must **justify** your decisions and also show that you have based your design choices upon a good knowledge of **existing** alternative designs and methods currently available.

Sample response extract

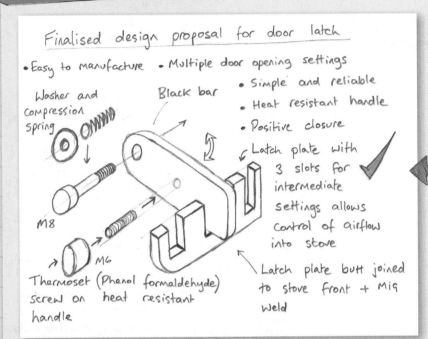

Finalised design proposal for door latch

- Easy to manufacture • Multiple door opening settings
- Washer and Compression Spring
- Black bar
  - Simple and reliable
  - Heat resistant handle
  - Positive closure
  - Latch plate with 3 slots for intermediate settings allows control of airflow into stove ✓

M8

Thermoset (Phenol formaldehyde) screw on heat resistant handle

M6

Latch plate butt joined to stove front + MIG weld

> This is the chosen design for the stove door latch from page 166, which would need to be fully justified. In this example, it could be shown using research that the decision is based upon knowledge of currently available products. Images of existing stove door latches that have been previously analysed could be shown.

> This extract from a response shows just one element of the overall stove design. You will need to **justify choices** for every single aspect of a developed final proposal.

Sample response extract

I have chosen the stove door latch design shown above because it fulfils the requirements of the design brief better than other designs I have explored.

1 It is straightforward and cost effective to manufacture.

2 It protects the user from burnt fingers.

3 Multiple door opening positions help to control airflow.

4 It will not jam if the parts expand from the stove heat.

5 Both major parts can be laser cut accurately in large quantities, for mass production of the product.

6 The resistance of the latch can be adjusted using the spring-assisted M8 socket head cap screw.

7 Ancillary parts can be stainless steel to enhance aesthetics and longevity.

> This extract represents only a small part of the overall justification of the proposal. For your chosen design you will also need to justify in detail the **material selection**, **manufacturing processes**, **sustainability**, and **information** to enable a competent engineer to manufacture the proposal accurately (see pages 168–169).

Now try this

The image shows a simple bottle opener designed in stainless steel. The client requested a simple one-piece design that was easily stored.

Justify the choice of the final design for this product.

Justifying materials and processes

When choosing the material a developed design is to be made from, you must show that you have based your choice on information and knowledge gained from thorough research. This also applies to your decisions about how to manufacture a product.

The decision process

The diagram below shows the process you should work through when reaching decisions about materials and manufacturing methods for your design.

1 Analyse product requirements

2 Research suitable materials

3 Analyse material properties

4 Compare advantages and disadvantages

5 Make decision and justify choice

6 Research possible manufacturing methods

7 Compare advantages and disadvantages

8 Make decision and justify choice

Sample response extract

1 A bicycle frame needs to be light and fairly stiff.

2 Possible materials for the cycle frame could be:

Titanium

OR

Carbon fibre

3A Highest strength to weight ratio of any metal, highly resistant to corrosion, very stiff, very expensive.

3B Highest tensile strength but brittle, high stiffness, good strength to weight ratio, corrosion resistant, if damaged cannot be repaired easily.

4 Very similar properties. Both very light and strong. Both corrosion resistant, and very stiff. Both excellent for bicycle frames.

5 Difficult choice and I intend to consider manufacturing processes before finalising my decision.

6 Titanium tubes cut on power hacksaw, clean oxide with carbide deburring tool, clean with acetone. Hold in frame jig and TIG weld. **Carbon fibre** and resin substrate into pre-set mould. Inflatable bladder inside tubes. Bake in enclave until cured.

7 Both materials have the qualities and properties for the product but carbon fibre can fail catastrophically and production is costly.

8 Carbon fibre manufacture is too costly for this one-off prototype. Titanium is more suitable for single item production.

This extract from a response shows the material and manufacturing choices in response to a brief for a prototype bicycle frame. It looks at some basic properties and manufacturing methods and is a good basis for expanding upon in a development process. It is important to suggest alternatives for manfacturing processes and types of material.

Now try this

Aluminium could also be considered as a suitable material for the bicycle frame in the example above. Identify the main properties of aluminium and suggest whether it is suitable for a bicycle frame.

Developing sustainability

As an engineer, you need to consider seriously the sustainability of a product at all stages of its life cycle. The example below shows sustainability as a major consideration and eight ways a product has been designed to be more sustainable, using technical terminology. Although it may not be possible to always fulfil the eight criteria, it is important to fulfil as many as possible.

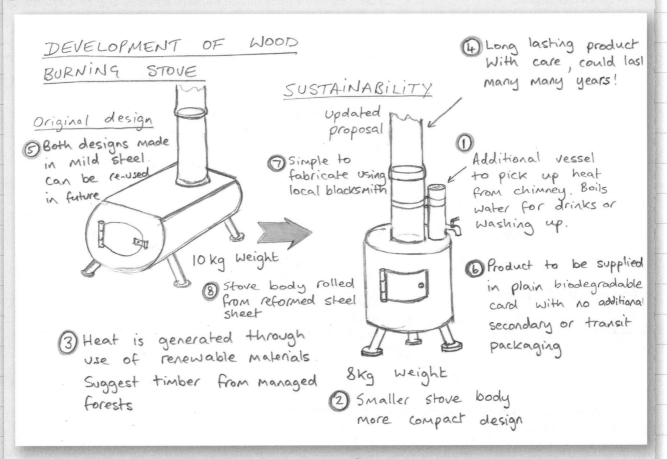

DEVELOPMENT OF WOOD BURNING STOVE

Original design

⑤ Both designs made in mild steel. Can be re-used in future.

10 kg Weight

③ Heat is generated through use of renewable materials. Suggest timber from managed forests

SUSTAINABILITY

Updated proposal

⑦ Simple to fabricate using local blacksmith

⑧ Stove body rolled from reformed steel sheet

8kg weight

② Smaller stove body more compact design

④ Long lasting product With care, could last many many years!

① Additional vessel to pick up heat from chimney. Boils water for drinks or washing up.

⑥ Product to be supplied in plain biodegradable card with no additional secondary or transit packaging

① I have added a secondary purpose: a vessel to collect radiated heat from the chimney to boil water.

② The design has been made smaller and less wasteful on materials.

③ Natural energy sources are used.

④ Designed for longevity.

⑤ Fabricated in mild steel, easily be recycled.

⑥ Biodegradable packaging.

⑦ Produced locally.

⑧ Reuse of material.

Now try this

Use the sample response extract to help you consider the different factors.

The luminAID is a solar-powered light for use in developing countries. It packs up very small and provides 16 hours of light from 7 hours of charging in sunlight.

Consider the luminAID and the images opposite. Give two ways that the manufacturer of this product made it as sustainable as possible.

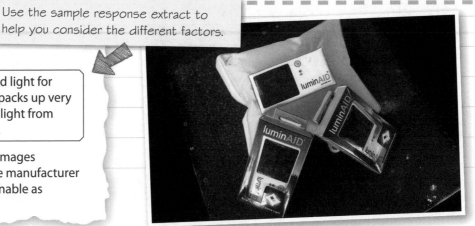

Validating a design proposal

Validating a **design** is about ensuring that the final product conforms to client specifications and requirements. It represents the final stage in the process of making sure that new designs are **fit** for the intended purpose, and that they **fulfil** the initial brief.

For example, you should evaluate your modified design for:

- success and limitations of the complete solutions
- indirect benefits and opportunities
- constraints
- opportunities for technology-led modifications.

Show your skills

Consider how you will give a balanced and thorough appraisal of:

- successes and opportunities and limitations and constraints of the completed solutions
- indirect benefits and opportunities.

You should also:

- provide a sound **rationale** for why the design solution is more effective in relation to the brief
- communicate further **technology-led** modifications, with detailed evidence of how they could optimise the solution. Be sure to make specific comments that relate to your chosen design solution, rather than general comments.

Evaluating a design proposal

Start	⟷	Finish
Product design specification (PDS)		Final design proposal

Appraise your design against the PDS.
Success?
Limitations?

Balance **benefits** and **opportunities** against **constraints**.

Discuss whether further **technological advancements** could optimise the solution.

Give a sound rationale for why your design is effective in relation to the brief.

To show that you have thoroughly **analysed** your final design proposal against the Client Brief and Product Design Specification, you could split your validation into the four sections shown here.

Sample response extract

✓

Specification point from PDS	Design 1	Design 2	Design 3	Design 4	Design 5
Ease of operation	0	0	3	0	4
Cost effectiveness	0	0	3	0	0
Ease of manufacture	1	1	3	0	1
Durability	1	1	0	3	0
Sustainability	1	0	1	4	1
Aesthetic quality	0	0	2	1	0
Total	3	2	12	8	6

In this extract from a validation exercise, each design has been **rated** against points from the PDS. Each design idea has been given a numerical value to establish the one that **matches** the PDS best. When a **numerical analysis** is completed you would need to provide a **justification** to support the decisions. This would include details of any **practical testing** of your design proposal you have done.

Now try this

Look at the development of the door handle/latch for a wood-burning stove, on pages 165–167 and page 169. Give four valid reasons why the final design proposal is an improvement over the initial simple latch design and the sliding bar design.

Some criteria for the optimised design are listed on page 167.

Evaluating with tools and techniques

There are many different **tools** and **techniques** you can use when evaluating the suitability of a design proposal or validating your choice of design.

Sample response extract

Evaluation

Criteria from P.D.S	Simple Latch	Sliding Latch	Multi-Position
Safety	0	2	5
Holds door partially open	0	0	5
Ease of manufacture	5	1	3
Reliable Operation	5	4	5
Cost Effective	5	3	3
TOTAL	15 ✗	10 ✗	21 ✓

CHOSEN LATCH DESIGN

Compression Spring
Black bar
Washer
M8
M6
Thermoset handle
Latch plate
Butt Weld to stove front

COST OF MANUFACTURE
① All three proposals use mild steel in roughly equal amounts. The sliding latch is too costly to consider. The multi-position latch does need slots cutting in the latch housing, but does not require bending like the 'simple latch' The manufacturing cost of these is likely to be similar

② This design can easily be dismantled for cleaning/maintenance

③ Peer survey

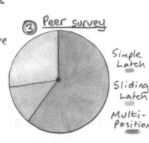

Simple Latch
Sliding Latch
Multi-Position

④ RATIONALE
The multi-position latch is more difficult to make than the simple latch, but it is much safer for the user, because it has a heat resistant handle. It also allows control of air into the stove by providing multiple positions for door opening.
Manufacture could be made easier if laser cutting facilities were available.

⑤ FURTHER DEVELOPMENT ?
The longevity of this design may be compromised because the heat of the stove may affect the heat treatment of the spring? It may be possible to consider an alternative method
'Wing nut' tensioner ?

You should use a variety of good **evaluation tools** in your validation. Notice in this extract:

① weighted matrix

② indirect benefit identified

③ graphical representation of survey results

④ a written rationale

⑤ identification of further opportunities (make sure you also consider further technology-led modifications, in relation to the specified product and solution).

Now try this

Look at the stove design on page 165. The feet are welded to the four tubular legs, which are then welded directly to the body of the stove.

Evaluate this design for the legs, giving two positive features and two negative features.

Think about the stove in operation, and consider if the proposed design would be a perfect solution.

Microcontrollers and Unit 6

In this unit, you will revise how to program or code microcontrollers to solve engineering problems.

What is a microcontroller?

A microcontroller:

- contains all the internal components of a computer on a single chip
- runs a stored computer program that controls a system.

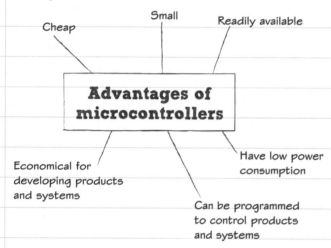

Cheap Small Readily available

Advantages of microcontrollers

Economical for developing products and systems

Have low power consumption

Can be programmed to control products and systems

Hardware and software

The only acceptable microcontroller hardware and programming languages for Unit 6 are shown in the table below.

Hardware device family	Software Integrated Development Environment (IDE)	Programming language
Arduino™/ Genuino™	Arduino™ IDE or Flowcode	Arduino™ C or Flowchart
PIC®	MPLAB® IDE (MPLAB® C) or Flowcode	C or Flowchart
PICAXE®	PICAXE® Editor	BASIC or Flowchart
GENIE®	GENIE® Studio or Circuit Wizard	BASIC or Flowchart

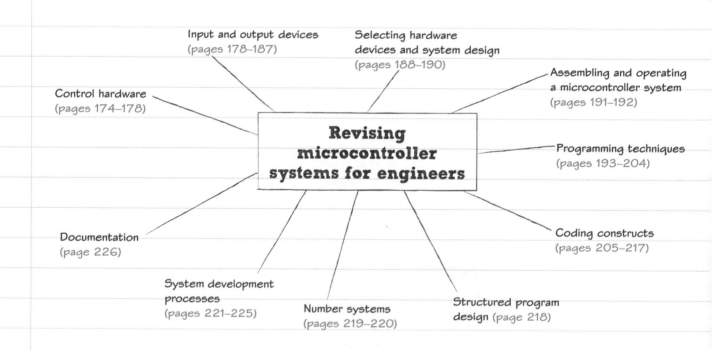

Input and output devices (pages 178–187)

Selecting hardware devices and system design (pages 188–190)

Assembling and operating a microcontroller system (pages 191–192)

Control hardware (pages 174–178)

Revising microcontroller systems for engineers

Programming techniques (pages 193–204)

Documentation (page 226)

Coding constructs (pages 205–217)

System development processes (pages 221–225)

Number systems (pages 219–220)

Structured program design (page 218)

Now try this

Identify any priority areas for your revision.

You can also refer to the following documents, which are kept up to date on the BTEC Nationals Engineering page of the Pearson website:

- Unit 6: Microcontroller Systems for Engineers unit specification
- Unit 6: Sample Assessment Material (SAM).

Comparing different microcontrollers

Microcontrollers are available in a **range of specifications**. The table below shows some important properties of one typical microcontroller from each family.

② The **processor speed** indicates how fast the system will operate.

③ Program **memory** is space available to store the software that determines how the microcontroller operates.

④ **RAM** is used to store the **data** used by the program.

⑤ **Interrupts** allow the normal operation of the program to be **altered** in response to an event.

⑥ The **stack** determines the **sequence** in which some parts of a program operate.

⑦ Each of the microcontrollers considered by this unit is available in a **range** of specifications.

⑧ **Inputs** allow devices, such as switches, to interact with the microcontroller.

⑨ **Outputs** allow devices, such as light-emitting diodes, to be turned on and off by the microcontroller.

① The **processor** is the primary component of the microcontroller. For the Genuino™ family a range of different processors are used. The other microcontrollers are based on specific types of processor.

Hardware family	Genuino™	PICAXE®	GENIE®	PIC®
1 Processor	ATmega328P	PICAXE18M2	Genie 18	PIC18F14K22
2 Processor speed	16	32	16/32	16 / 31
3 Program Memory (kilobytes)	32	2	10	16
4 Random Access Memory (RAM)	2048	512	256	512
5 Interrupts	2	8	8	10
6 Stack	Unlimited	8	32	31
7 Typical processor cost	£2.39	£2.77	£1.36	£1.61
8 Digital input capability	Up to 14	Up to 16	6	Up to 17
9 Digital output capability	Up to 14	Up to 16	9	Up to 17
10 Analogue to digital inputs	6	10	3	12
11 PWM outputs	6	2	8	4
12 Project board	Arduino™ Uno Board	PICAXE® CHI-030 Project Board	Genie® PCB118 E18 Activity Kit	DM164130-9-PICkit Low Pin Count Demo Board

⑩ When the microcontroller needs to connect to sensors that measure analogue variables (such as temperature), **analogue to digital conversion inputs** need to be used.

⑪ **PWM** (Pulse Width Modulation) allows the digital microcontroller to **simulate an analogue output**.

⑫ A microcontroller processor needs to **connect** to other components in order to become a functional system. The most common method is to insert the microcontroller into a **project board** that contains the other components needed. Boards are available in a range of specifications to suit the microcontroller and its application.

Now try this

Using a detailed specification from the internet for one microcontroller, revise the important properties.

Project boards

You can use microcontroller project boards to easily **connect** the microcontroller to **input** and **output** devices. Some project boards allow the microcontroller to be programmed on them.

Project boards

Breadboards provide a quick way to **test** a temporary **system**.

- The project boards shown here provide one example for each of the microcontroller platforms. There are many others available. The different styles of board reflect the different uses they are intended for.

- At one end of the range, the GENIE® board forms a permanent microcontroller system where the user connects their own input output devices. At the other end of the range, the E-blocks system is intended to be used as part of a training package.

GENIE® **project boards** are supplied as kits of components that you can solder together. Input and output devices are then soldered to the project board to make the system complete.

Arduino™ **project boards** are supplied assembled, ready to use. Wires are inserted into the header sockets on the microprocessor and connected to components on the breadboard.

The PICAXE® **experimenter board** includes a range of input / output devices built onto the board. You can connect them to the microcontroller via the breadboard using wires.

The **E-blocks system** is made up of two components, the programmer and the project board. This development board provides an extensive range of input/output devices. E-blocks are intended to be used as part of a training package. They would not be built into a working system.

The **PICkit 3 system** consists of the programmer and a project board.

Now try this

You are designing a device to control an alarm system. List the advantages of using a project board for one of the microcontroller platforms.

Using flowcharts

When designing a system, you will use flowcharts to produce a microcontroller program by **placing** and **connecting symbols**.

Common symbols used in a flowchart

The shapes of the symbols used in flowcharts are specified by BS 4058:1987, ISO 5807:1985. The shapes and meanings of some of the most common ones are shown.

Terminator – this indicates the start or stop of the program.

Process – a process or operation is carried out.

Call – directs the flow of the program to a separate subroutine.

Decision – perform one of two actions depending on the result of the decision.

Input or output – of data.

Flow lines – these control the program sequence.

Example of flowchart

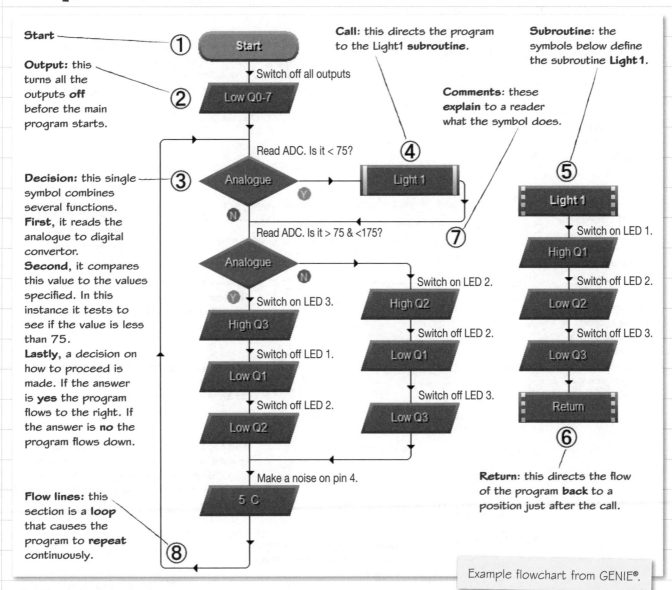

Start

Output: this turns all the outputs **off** before the main program starts.

Decision: this single symbol combines several functions.
First, it reads the analogue to digital convertor.
Second, it compares this value to the values specified. In this instance it tests to see if the value is less than 75.
Lastly, a decision on how to proceed is made. If the answer is **yes** the program flows to the right. If the answer is **no** the program flows down.

Flow lines: this section is a **loop** that causes the program to repeat continuously.

Call: this directs the program to the Light1 **subroutine**.

Comments: these **explain** to a reader what the symbol does.

Subroutine: the symbols below define the subroutine **Light1**.

Return: this directs the flow of the program **back** to a position just after the call.

Start
Switch off all outputs
Low Q0-7
Read ADC. Is it < 75?
Analogue
Light 1
Read ADC. Is it > 75 & <175?
Analogue
Switch on LED 2.
High Q2
Switch on LED 3.
High Q3
Switch off LED 2.
Low Q1
Switch off LED 1.
Low Q1
Switch off LED 3.
Low Q3
Switch off LED 2.
Low Q2
Make a noise on pin 4.
5 C
Light 1
Switch on LED 1.
High Q1
Switch off LED 2.
Low Q2
Switch off LED 3.
Low Q3
Return

Example flowchart from GENIE®.

Now try this

Record other flowchart symbols that are used to write microcontroller software.

BASIC as a programming language

BASIC stands for **B**eginner's **A**ll-purpose **S**ymbolic **I**nstruction **C**ode. You can use BASIC to write software even if you are not an expert programmer.

BASIC programs

BASIC programs consist of **five** parts:

1 **Commands** – the key words or names given to the tasks the microcontroller can perform.

2 **Statements** – lines of code that contain a complete instruction for the microcontroller.

3 **Subroutines** – sections of code that can be reused in different parts of a program.

4 **Declarations** – statements that name a variable or constant.

5 **Operators** – used to perform mathematical or logical functions, or comparisons. For example, adding two numbers, or testing if one value is less than another.

Example of a BASIC program

Constants are numbers, or ASCII text strings, that **do not change** throughout the program.

Comments allow the writer to **provide information** about the program. A comment starts with the semicolon (;) or apostrophe (') character, and does not affect the way the program operates. Comments are coloured **green**.

The **If** command allows the program to **complete different tasks depending on the result** of the condition being tested. In this example, a comparison between a variable and a number constant is considered.

```
;This is an example of a BASIC program
;It will be used to highlight common features of the language   ②

main:
      readadc c.1,b1          ' read the voltage on pin 18 into variable b1
                      ①
   ③ if b1<75 then light1     ' if b1 is less than 75 then light 1
            ⑧
      if b1<175 then light2   ' if b1 is less than 175 then light 2
      goto light3             ' if b1 is greater than 175 then light 3

light1:
    ⑤ high 1                  ' switch on LED 1
      low 2                   ' switch off LED 2
      low 3                   ' switch off LED 3
      goto beep              ' make a noise
      goto main              ' loop

light2: ⑦
      low 1                   ' switch off LED 1
      high 2                  ' switch on LED 2
      low 3                   ' switch off LED 3
      goto beep              ' make a noise
      goto main              ' loop

light3:              ④
      low 1                   ' switch off LED 1
      low 2                   ' switch off LED 2
      high 3                  ' switch on LED 3
      goto beep ⑥            ' make a noise
      goto main ⑥            ' loop

beep: ⑥
      Sound B.4,(55,25)
      Goto main
```

Variables, such as 'b1', are named sections of **RAM memory** where temporary data can be stored.

Commands, such as 'high', are **instructions** that make the microcontroller perform a **specific task**, such as setting the output on a specific pin to be high or low.

Labels, such as 'light 2', are used to mark a **particular place** in a program. The program can then go to these places when directed.

Whitespace is a method of **inserting blank spaces** into a program to make it easier to read. For example, adding blank lines and indenting commands helps identify groups of commands.

Beep is a **subroutine** that is used by **other parts** of the program.

BASIC programming language.

Now try this

List the parts of a BASIC program.

C as a programming language

C is a structured programming language that is widely used to write computer software. One of the reasons for its popularity is the **efficient machine code** it produces.

Parts of a C program

C programs comprise three main parts: structure, values and functions.

1 The **structure** of a C program refers to the way in which the parts link with each other. The structure must follow the rules of the language.

2 **Values** include constants and data types.

3 **Functions** are the components of the program that perform tasks.

Example of a C program

Comments allow the writer to **provide information** about the program. Comments **start with //**, and do not affect the way the program operates.

void setup() is a **function** that only runs when the program starts. It is used to **initialise the factors**, such as variables and the configuration of the microcontroller.

int tells the program that the following named **variable** is of an **integer type**. This means it will be a whole number in the range of −32768 to 32767. In this example the variable lowValue is assigned the number 75.

32 768 = 2^15

for repeats a **block of code a number of times**. In this example the variable pinNumber starts at 2 and increases by 1 until it reaches 5. The next section of code sets the pin with the same value as pinNumber to an output and turns it off. For example, when pinNumber = 2, pin 2 is set to be an output and turned off.

```
① // This is an example of C program
   //It will be used to highlight common features of the language

② void setup() {
       // set the LED pins as outputs, the for() loop reduces the number of lines needed
③  for (int pinNumber = 2; pinNumber < 5; pinNumber++) {
④      pinMode(pinNumber, OUTPUT);
⑤      digitalWrite(pinNumber, LOW);
   }
   }
   void loop() {
   // This reads the ADC on pin A0 and assigns the value to the integer variable called sensorVal
       int sensorVal = analogRead(A0); ⑥
⑦     int lowValue = 75;            // This defines the integer value below which LED 2 will turn on
       int middleValue =175;        // This defines the integer value above which LED 4 will turn on

   // This tests to see if the ADC value is less than 75
⑧ if (sensorVal < lowValue) {
       digitalWrite(2, HIGH);        // This turns on LED 2
       digitalWrite(3, LOW);         // This turns off LED 3
       digitalWrite(4, LOW);         // This turns off LED 4
   }

   // This tests to see if the ADC value is equal to or more than 75 and less than 175
   else if (sensorVal >= lowValue && sensorVal < middleValue) {
       digitalWrite(2, LOW);         // This turns off LED 2
       digitalWrite(3, HIGH);        // This turns on LED 3
       digitalWrite(4, LOW);         // This turns off LED 4
   }

   // This tests to see if the ADC value is equal to more than 175
   else if (sensorVal >= middleValue) {
       digitalWrite(2, LOW);         // This turns off LED 2
       digitalWrite(3, LOW);         // This turns off LED 3
       digitalWrite(4, HIGH);        // This turns on LED 4
   }

⑨ tone (5,440,250);
   }
```

pinMode is the function that **sets the microcontroller pin** to be an **input** or **output**.

digitalWrite is the function that **turns a pin** either **on** or **off**.

analogRead is the **function** that **reads the value** of an **analogue to digital converter** (ADC). In this program the value is read for ADC on pin A0, and is assigned to the variable called sensorVal.

if is a **control structure** that is used to **compare two values** and take a particular **action** based on the **result** of the comparison.

tone causes a **sound to be produced** on an **output pin**. In this example a note of 440 Hz is generated on pin 5 for 250 ms.

C program for Arduino™.

Now try this

Rewrite, on paper or on a computer, parts of the program to work to the following new specification:
The ADC is on pin A2. The value below which LED 2 turns on is 50 and the value above which LED 4 turns on is 200. The tone should be a frequency of 200 Hz, generated on pin 6 for 100 ms.

Input and output devices

Input and output devices allow microcontroller systems to **react to**, and **change**, the **environment in which they operate**.

Analogue inputs and outputs

These are inputs or outputs that are **continuously variable**.

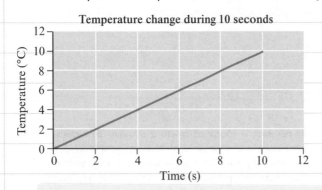

The graphs show how a temperature changes over time. An increase in time corresponds to an increase in temperature. If you could measure time to an accuracy of 10^{-999} the corresponding temperature would also be measureable with the same accuracy.

Digital inputs and outputs

These are inputs or outputs that have a **fixed range of values**.

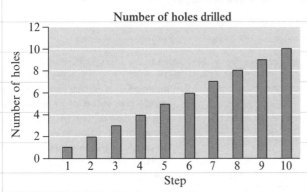

The graph shows the number of holes drilled per step of a manufacturing operation. The number of holes always increases by 1, not by fractional amounts. Either there is a hole, or there is not.

Modular input/output devices

Example of a modular sensor.

Allow you to assemble more complex systems that would not be possible using just discrete components.

Modular devices

Combine all the components needed to connect directly to the microcontroller system.

You only need to know how to get signals from, or to, the device.

Input and output devices

- **Input devices** allow microcontroller systems to react to the environment.
- Input devices may measure, or react to, analogue or digital signals.
- The data from analogue sensors must be converted into a digital format in order for microcontrollers to process the data.
- **Output devices** cause a change in the environment the microcontroller operates in.

Discrete input output devices

A discrete input or output device is one that consists of a single component.

Advantages	Disadvantages
low cost	may need additional components connected to it
flexible – many applications	may require complex circuits connected to them

Now try this

Identify where sensors are used in your home; for example, in a heating system. Describe what the sensor measures and what the output is as a result of this measurement.

Switches

The switches revised on this page can be used to provide a microcontroller with a **digital input signal**.

Switches and buttons

Switches can have either a toggle or momentary action.

- **Toggle** means that the switch will stay in whichever position it is moved to, on or off.
- **Momentary** means that the switch changes from on to off only while it is being acted on by the user. They are either normally on, or normally off.

Switches are available with a wide range of options for the number of connections they make or break. The rotary switch shown opposite has 12 switches that turn on in sequence when the shaft is turned.

Slide, Rocker, Push, Toggle, Pull, Rotary

Switches and buttons are operated by a person. The major difference in these switches is the **type of movement** that has to take place for the switch to change from being on to off.

Movement and orientation switches

Tilt switches and shock switches can be mounted in a horizontal or vertical position.

- **Tilt** switches can be used as safety devices; for example, turning off heaters if they fall over.
- **Shock** switches react when a movement occurs at 90° to their body of the switch.

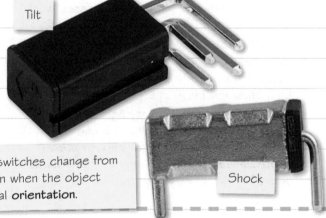

Tilt, Shock

These **tilt** or **shock** switches change from on to off, or off to on when the object moves from its normal **orientation**.

Micro-switches

Depending on how a micro-switch is configured, it may react when two objects come together or when they move apart.

- Typical uses of micro-switches include **triggers** for alarms and safety **interlocks** for guards on machines.
- The **lever arms** on the top of the micro-switch are available in a wide range of styles and sizes. This is to allow the switch to be mounted in an appropriate location while still being able to react to movement.

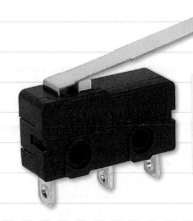

Micro-switches are designed to detect the **relative movement** between two objects.

Now try this

Identify a range of situations where tilt and micro-switches are used.

Visible and infrared light-sensing devices

The sensors described in this page allow a microcontroller to **react to**, or **measure**, **changes in the intensity** of visible or infrared light.

Light-dependent resistors

Light-dependent resistors (LDRs) change their **resistance** depending on the **intensity** of light that hits their surface.

- **In the dark**, LDRs typically have a resistance of $1\,M\Omega$.

- **In bright light** their resistance may fall to around $400\,\Omega$.

- It may take several seconds for an LDR's **resistance to change** in **response** to light intensity.

An ORP12 LDR. An **LDR** can be read by a microcontroller by converting the resistance to a voltage reading.

Phototransistors and photodiodes

Phototransistors and photodiodes **react** much more **quickly** than LDRs to changes in light.

- A **photodiode** works in the same way as a solar cell. Light that hits the sensor generates a small current. A photodiode will work in 1 nanosecond or 10^{-9} s.

- **Phototransistors** work by light hitting the sensor, which then allows a current to flow between the collector and emitter. A phototransistor will typically switch between on and off in 10 microseconds or 10^{-6} s.

Phototransistors and **photodiodes** can be used to detect changes in the intensity of both visible and infrared light.

Infrared receivers

Data is transmitted to an infrared receiver by adding a code to a carrier signal, in the same way as a radio is broadcast.

The receiver's circuitry removes the carrier signal, leaving only the code for the microprocessor to process.

Format

The exact **format of the codes** is specified by a **range of protocols**. Most protocols transmit the data by sending specific sequences of high and low signals.

Infrared receivers are optical sensors that receive signals transmitted by devices such as television remote controls.

Now try this

Think about devices, or objects, that make use of visible light sensors; for example, street lighting. Describe what happens when the level of light changes and explain why this is an advantage for the user.

Temperature and humidity sensors

The sensors described on this page can be used to turn **on** or **off** devices connected to a microcontroller, such as a fan. They can provide **information** to be displayed to a user on devices such as LCD screens.

Thermistors

- The **resistance** of a thermistor either increases (often used to protect circuits) or decreases (often used to measure temperature) with a rise in temperature.
- Accurate measurement using a thermistor can be complex, as the **resistance change** is not linear.
- Thermistors typically operate in a **temperature range** of –50 °C to 300 °C.

A **thermistor** and a **BASIC program** to turn on an output dependent on temperature.

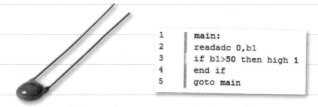

```
1    main:
2    readadc 0,b1
3    if b1>50 then high 1
4    end if
5    goto main
```

Temperature sensors

- This temperature **sensor** is an integrated circuit that outputs a voltage, which is **directly proportional** to the temperature.
- A rise of 1 °C will increase the output voltage by 10 mV. This **predictable performance** makes it easier to accurately measure temperature using a microcontroller.

A **temperature sensor** to turn on an output dependent on the temperature. This sensor operates in a temperature range of –40 °C to 125 °C. It can be utilised using the **BASIC program** above.

Environmental sensors

- **Temperature** and **relative humidity** can be measured with a sensor such as the DHT11 shown here, which produces a **digital output signal** that can be displayed on a computer monitor or an LCD screen.
- It measures **relative humidity** between 20–80% and temperatures with a range of 0–50 °C.

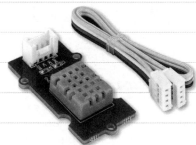

A **DHT11 sensor** and **code** for the Arduino™ Uno to read and display the values.

```
void loop() {
  // The next lines read the sensors humidity and temperature value
  float h = dht.readHumidity();
  float t = dht.readTemperature();

  // The following lines will display the readings from the sensor
  if (isnan(t) || isnan(h)) {
    Serial.println("Failed to read from DHT");
  } else {
    Serial.print("Humidity: ");
    Serial.print(h);
    Serial.print(" %\t");
    Serial.print("Temperature: ");
    Serial.print(t);
    Serial.println(" *C");
  }
}
```

Now try this

Using the internet, locate specifications of the components shown above to aid your revision and provide reference.

Input interfacing requirements

You will need to consider several factors when **connecting sensors** to **microcontrollers**.

Signal conditioning

This is the process of ensuring that the output signal of a sensor matches the input requirements of a microcontroller. **Requirements include:**

- **Isolation** – this is a barrier between the sensor and microcontroller to protect it from damage.
- **Amplification** – this increases the voltage output by the sensor.
- **Filtering** – this is the process of removing unwanted information, possibly from interference, from a signal.

When an analogue signal is converted to digital the signal is assigned the nearest digital value.

Analogue-to-digital conversion

Analogue signals can take **any value**; for example, it could be any number between 1 and 1.1. **Digital signals** have a **limited set of values**; for example, 1 or 1.1 with no values in between these.

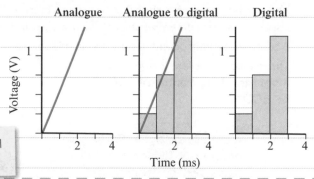

Modular sensor boards

Modular sensor boards:

- provide a **ready assembled sensor** complete with all the required interfacing circuits
- allow the user to focus on **developing the programs** to use the sensors rather than constructing the electronics needed to turn the basic sensor into a working device.

Links See the ultrasonic sensor on page 183 to revise an example of a modular sensor.

Pulse width modulation (PWM)

PWM is a method of **encoding a signal** by **varying** the length of time a pulse of a set length is on or off.

This shows an **equal** pulse length.

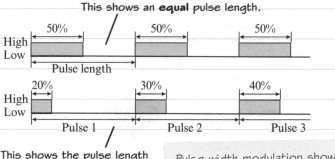

This shows the pulse length **changing** over 3 pulses.

Pulse width modulation showing several PWM signals.

Serial communication

Serial communication is a method of **encoding data** by sending a series of high or low voltages in a **pattern**. Every piece of data sent is made up of the same number of bits of information, only the sequence of high and lows will change.

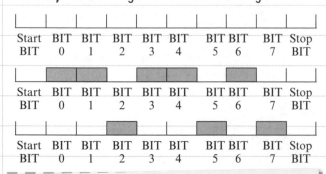

Inter-integrated circuit (I²C)

Serial communication requires **multiple wires** to transmit a signal into a microcontroller but an I²C **connection** only **requires two**.

- One of the wires sends a clock or timing signal.
- The other wire sends the data signal.
- The relative levels of the voltages between the clock and the data signal, and when they change, are used to control what type of signal is being sent.
- An I²C allows multiple devices to be connected.
- The first part of a signal is an address that determines where the data are to be sent to. The second part of the signal is the data itself.

Serial signals – The **first** image shows the set length of eight data bits, plus one bit to indicate the start and one for the stop of each pattern. The **second and third images** show examples of data being received.

Now try this

Many types of entertainment media, such as mp3 files, are provided in a digital format. List other types of entertainment media that use digital data.

Ultrasonic and control potentiometers

Ultrasonic and control potentiometers allow a microcontroller to **react to**, or **measure**, **distance** and **position**. The operation of these devices is described below.

Ultrasonic sensors

Ultrasonic sensors use **sound** to determine the **distance** between the sensor and objects.

- A **high frequency sound**, above the range that can he heard by humans, is sent out by the **ultrasonic transmitter**.
- This sound then **bounces back** from any object in front of the transmitter and is **detected** by the **ultrasonic receiver**.
- The time taken for the sound to travel from the transmitter to the receiver is used to **determine the distance**.

Ultrasonic sensor with program.

```
1    symbol SIG = C.1 ; This specifies which pin the sensor is connected to
2    symbol range = w1 ; This creates a place to store the value from the sensor
3    main:
4    ultra SIG,range ; This reads the sensor and stores the value in the place called range
5    debug range ; This causes the value from range to be displayed on a computer monitor
6    pause 50 ; This is a short delay
7    goto main ; This causes the program to repeat
```

Control potentiometers

Control potentiometers are used to provide a microcontroller with **information** about a **position set** by a user.
The user can either **rotate** or **slide** the potentiometer to a **chosen position**; for example, to adjust the volume of an audio amplifier.

- The **output voltage** can be **adjusted** between the supply voltage and 0 V, varying with either a linear or logarithmic scale.

Logarithmic scales

A **logarithmic scale** increases in orders of magnitude; for example, each mark on a scale might be 10 times the previous mark. They are useful for audio controls, as humans perceive relative volume levels in an approximate logarithmic scale.

Rotary and **slide control potentiometers**. These can be read by a microcontroller using the analogue to digital command.

Now try this

Consider how rotary and slide potentiometers are used. Compare their relative advantages and disadvantages.

Optoelectronic output devices

Optoelectronics use electricity to **generate light** in a variety of ways. Some of the most useful types of optoelectronics are described below.

Light-emitting diode (LED)

LEDs can be turned on and off by setting microcontroller output pins to high or low. They are available in two main types.

1 Those that emit **visible light**. Visible light LEDs are often used as indicators.

2 Those that emit **infrared light**. Infrared LEDs are combined with infrared receivers to transmit data, as used, for example, in TV remote controls.

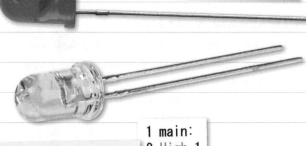

```
1 main:
2 High 1
3 Stop
```

Visible and infrared light-emitting diodes, with example of BASIC code to turn on a LED.

7-segment display

7-segment displays consist of **seven rectangular LEDs** arranged in the pattern shown opposite. By turning on and off different combinations of the LEDS, the numbers 0–9 can be displayed. For example, the number 6 is displayed as shown.

7-segment displays can be controlled by attaching them to seven output pins of a microcontroller, or by connecting via a Binary Coded Decimal (BCD) driver, such as a 7447 chip. Using a BCD driver reduces the number of control pins needed from seven to four.

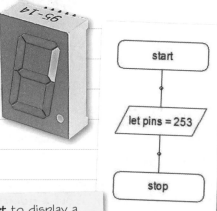

Example flowchart to display a number on a **7-segment display**.

Liquid crystal display (LCD)

LCDs display **text** and **numbers on thin plastic displays**.

- The numbers and text can scroll along the screen, or they can remain stationary.
- LCD screens are available in a range of colours and may include a backlight.
- LCDs are controlled by connecting them to the output pins of the microcontroller. Typically, six pins are used to display the full range of characters available.

```
void setup() {
  // set up the LCD's columns and rows:
  lcd.begin (16, 2);
  // Print a message to the LCD
  lcd.print("Winstar, WH1602L1");
}
```

Example C code to output text to a LCD.

Now try this

Investigate the methods used to generate a message on a LCD using a platform of your choice.

Electromechanical output devices

This page revises the **characteristics** and **operation** of **relays**, **motors** and **servos**.

Relay

In the relay, the low current from the microcontroller energises an electromagnetic coil.

⬇

This magnetism then pulls the armature towards the coil.

⬇

A moveable contact is attached to the armature.

⬇

The contact switches the electricity supply on or off.

A relay can be switched by a microcontroller by **connecting** the coil to a digital output pin. A high signal will energise the coil.

A **relay** is a device that allows a low power signal to turn on or off a higher power electrical supply.

DC motor

A user can attach a wide range of devices to the motor's **shaft**, which will then **rotate**.

- Typically, a DC motor may rotate too quickly for the application it is being used in.
- Gears and pulleys are mechanical methods of controlling the speed, direction and torque of the rotation.

🔗 **Links**   Microcontrollers can be used to vary the speed at which the motor rotates by using the pulse width modulation (PWM) technique revised on page 182.

A **DC** (direct current) motor uses electricity to spin a shaft.

Servo

A servo is a device that uses an electric motor to **turn** a shaft through a **set number of degrees**. A common application of a servo is to control the position of objects attached to the shaft. Most servos have three wires connected to them. Two of the wires connect to a power supply and the third is the control signal. The duration of the electrical pulse controls the position of the servo shaft:

- A pulse width of 1.5 milliseconds (ms) will cause the servo shaft to go to the middle.
- A pulse width of less than 1.5 ms makes the servo rotate clockwise from the central position.
- A pulse width of more than 1.5 ms makes the servo rotate anticlockwise.
- Typically, pulse widths of 1 ms or 2 ms will cause the shaft to rotate to the ends of the range.

A typical **servo** rotates through 180° but some can rotate continuously.

The effect of different pulse widths on the position of the servo arm.

Now try this

Identify devices that use relays, DC motors or servos.

Audio output devices

Audio devices are used to make **sounds** when connected to a microcontroller.

Buzzers and sirens

Connecting a buzzer or siren to an output pin set to high will turn them on.

- **Buzzers** tend to operate at a single fixed frequency.
- **Sirens** tend to 'warble' between an upper and lower frequency and are typically much louder than buzzers.

> Buzzers and sirens are audio output devices that only require a power supply for them to generate a noise.

Piezo transducer

Piezo transducers of the type shown opposite are **small electronic devices**, typically less than 1 mm thick, that can be used to play sounds.

- Rather than simply turning them on as is the case with the buzzer, a piezo transducer must be supplied with a signal that generates the function required.
- Commands such as 'Sound' for PICAXE® and 'tone' for Arduino™ are used.
- Some microcontrollers, such as the PICAXE® and GENIE®, have sequences of sounds that play 'songs', prebuilt into them.
- Due to their low power requirements, piezo transducers can often be connected directly to a microcontroller.

> **Piezo transducers** can be fitted to greetings cards to play either music or speech. In order to generate a sound, the microcontroller must be told the **pin** that the transducer is connected to, the **frequency** of the sound and **how long** to play the sound.

Speaker

Most speakers use **electromagnetism** to move a cone of paper. The movement of the paper generates the sound that the user hears.

- Speakers are available in a wide range of sizes, typically from 5 cm to 450 cm.
- Large speakers usually need external amplification, as in most situations a microcontroller cannot supply enough current.
- The same commands as for the piezo transducer are used to generate the sounds.

> The size of a **speaker** affects the range of frequencies it can generate; the bigger the speaker, the lower the frequency that can be produced.

Now try this

List five examples where buzzers, sirens, piezo transducers or speakers are used.

Transistor output stages

A transistor is often used as an **interface** between a **microcontroller** and **output devices**.

The function of transistors

Most microcontrollers can supply only a relatively low current, typically around 20 mA per pin.

- An **interface** is needed if the microcontroller controls the operation of a device that takes a larger current, such as an electric motor.
- **Transistors** can work as an interface. A small current applied to one connection of the transistor can allow a much larger current to flow from another of the transistor's connections.

Transistor types

There are many different types of transistor.

- Transistors **vary** according to the speed at which they respond and the amount of current they can transmit. Generally, the faster the transistor can react then the smaller the current it can tolerate.
- Transistors that operate quickly with **low currents** are called **signal transistors**.
- Transistors that control **larger currents** are called **power transistors**.

Signal transistors

The BC238B is an NPN-type transistor. The three legs of this transistor are called the **base**, **emitter** and **collector**.

- The **base** connects to the microcontroller. It controls the output from the emitter.
- The **collector** is where current goes into the transistor.
- The **emitter** is where current flows out.

One typical signal transistor is the BC238B.

Transistor current gain

Transistor current gain is the relationship between the current flowing into the base of the transistor and the current flowing out of the emitter. For practical purposes this can be expressed as:

Gain = collector current/base current

- For the **BC238B** shown, the current gain is approximately 200. This means that a current of 1 mA flowing into the base will cause a current of 200 mA to flow from the emitter.
- A **Darlington pair** is where the emitter of one transistor is connected to the base of a second transistor. This multiplies the gain of each transistor. For the BC238B this would be 200 × 200, or a gain of 40 000.

Transistor interface circuits

This diagram shows a typical **transistor interface**. The diode, D1, is built into the circuit to stop the microcontroller being damaged by the output device when the power is turned off.

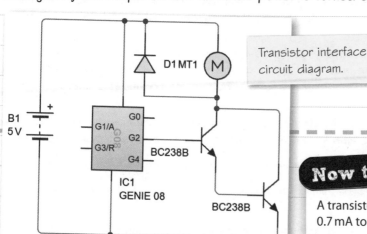

Transistor interface circuit diagram.

Darlington driver integrated circuits

As system designer, you might find it convenient to use an **integrated circuit** that combines the multiple transistors into a single component. One version of this type of driver is the ULN2003. This provides seven Darlington pairs in a single 16-pin integrated circuit.

Now try this

A transistor is required to increase the base current from 0.7 mA to 500 mA. Determine the gain required.

Selecting hardware: input devices

The table below indicates **typical parameters** associated with **input devices**. The information in the table is taken from a single supplier and gives a range of **characteristics** for each type of device. For output devices, see page 189.

Parameter to be measured	Device	Accuracy	Range	Importance of aesthetics	Control	Ergonomics	Cost from	Interface command
Digital	Switches	On or off	Not applicable	Possibly	User	Important	£0.08	Digital read
	Buttons	On or off	Not applicable	Possibly	User	Important	£0.17	Digital read
	Control potentiometer	Dependent on cost	470Ω to 1MΩ	Possibly	User	Important	£0.52	Analogue read
Temperature	Thermistor	Dependent on software	–40°C to 125°C	Not important	Environment	Not important	£0.30	Analogue read
	Temperature sensors	Dependent on software	–70°C to 500°C	Not important	Environment	Not important	£1.18	Analogue read
	Environmental sensor	Typically ± 2%	Temp 0°C to 50°C Humidity 20–90%	Not important	Environment	Not important	£3.87	Sensor specific
Light	LDR	At 10 lux 20–40kΩ At 100 lux 2 –4kΩ	Dark resistance 1MΩ Light resistance 2Ω	Not important	Environment	Not important	£0.44	Analogue read
	Infrared	Dependent on software	Up to 30m	Not important	Environment	Not important	£0.25	Digital or analogue read
Movement	Tilt switch	On or off	Detects 10° change in position	Not important	User or environment	Not important	£0.56	Digital read
Presence	Micro-switch	On or off	Not applicable	Possibly	User or environment	Not important	£0.11	Digital read
	Ultrasonic	Up to 0.3cm	2cm to 450cm	Not important	User or environment	Not important	£3.59	Length of pulse

Selection criteria

The selection of specific input devices depends on a wide range of factors.

- For each microprocessor control system there will be a different set of priorities, which will affect how you choose a particular device. As system designer, you will need to **balance** these sometimes conflicting factors. For example, the need for an accurate system may not match with the need for a low cost system.

- To provide the best solution you will need to **analyse** the key performance characteristics of the input devices. You will need to consider the interface between the sensor and the microcontroller. For example, the thermistor is the lowest cost temperature sensor but needs the most complex software to produce an accurate temperature measurement.

- When you have selected the most appropriate hardware devices for your system, you must show how the components are **connected** using diagrams that comply with appropriate international standards.

Now try this

Choose one of the components listed above and investigate the range of specifications available.

Selecting hardware: output devices

The table below indicates **typical parameters** associated with **output devices**. The information in the table is taken from a single supplier and gives a range of **characteristics** for each type of device. For input devices, see page 188.

Required output / effect	Device	Typical dimensions	Voltage	Current	Output feature torque, visibility or audibility	Cost	Interface command
Optoelectronic	Light-emitting diode	3 mm to 20 mm	Max 2.5 V	30 mA	Visibility – from 3 mcd to 4500 mcd	£0.07	Digital write
	7-segment display	8 mm to 45 mm high	Max 2.5 V	20 mA	Visibility – from 12 mcd to 100 mcd	£0.44	Digital write
	Liquid crystal display	From 52×32×9 mm	5 V	1.2 mA	Visibility depends on user control	£2.41	Digital write to multiple pins
Electromechanical	Relay	From 19×19×15 mm	6 V for control	60 mA for control	Able to switch up to 277 V AC and to 10 A	£0.56	Digital write
	Direct current motor	From 35 mm long, 20 mm diameter	3 V	From 34 mA	Torque – from 8.2 g cm	£0.54	Digital write or PWM
	Servo	From 23×12×29 mm	4.8 V	Not specified	Torque – 1800 g cm	£3.10	PWM
Audio	Buzzer	From 12 mm dia, 7 mm high	4–8 V	30 mA at 5 V	Audibility – resonance frequency 3.1 ± 0.5 kHz	£0.50	Digital write
	Siren	From 12 mm dia, 42 mm long	12 V	280 mA	Audibility – sound pressure level 105 dB at 12 VDC/100 cm	£5.82	Digital write
	Speaker	From 50 mm dia	2 V	250 mA	Audibility – resonance frequency 450 kHz, 86 dB at 1 kHz	£1.33	Frequency and amplitude
	Piezo transducer	From 12 mm dia, 1 mm high	12 V	2 mA	Audibility – 75 dB	£0.72	Square wave

Selection criteria

The selection of specific output devices depends on a wide range of factors.

- For each microprocessor control system there will be a different set of priorities, which will affect how you choose a particular device. As system designer, you will need to **balance** these sometimes conflicting factors. For example, the need for an accurate system may not match with the need for a low cost system.

- To provide the best solution you will need to **analyse** the key performance characteristics of the output devices. You will need to consider the power required to drive the device. For example, a siren may require 280 mA, whereas the maximum current available from a microcontroller output pin might be only 40 mA.

- When you have selected the most appropriate hardware devices for your system, you must show how the components are **connected** using diagrams that comply with appropriate international standards.

Now try this

Choose one of the components listed above and investigate the range of specifications available.

Generating a system design

Once you have selected **input** and **output devices** for your system design, you must decide how to **connect** them to the **control hardware**.

Schematic diagrams

A schematic drawing uses a range of **symbols** to represent the **input / output** components of the system. To make sure that all users interpret the drawing in the same way, the symbols used must comply with the set standards appropriate to the country the system will be used in. In the UK, for example, BS8888 and BS3939 set the standards to be used.

A range of **British Standard symbols** for common input and output devices.

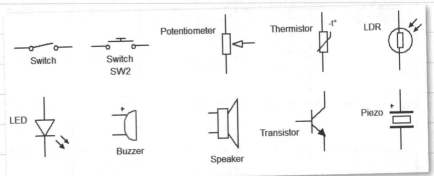

Circuit diagrams

To specify **connections** between input, output devices and control hardware the system designer draws **links** between each component.

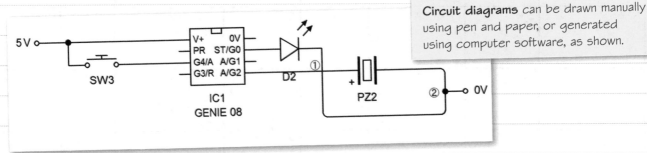

Circuit diagrams can be drawn manually using pen and paper, or generated using computer software, as shown.

Connections

When the system designer produces the circuit diagram the connections between components often **overlap**.

- The diagram needs to indicate if these overlapping lines represent **physical joins** between the connections.
- Where the **connections join**, a **dot** is drawn in the middle of the joint.

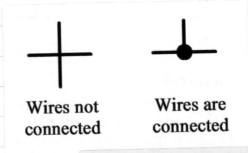

Wires not connected

Wires are connected

The approved convention of showing **overlapping wires** on a Computer-Aided Design (CAD) drawing.

Now try this

Identify the schematic symbols for at least five input and output components not shown above. You should ensure that the symbols conform to the appropriate international standard for your locality.

Safe use of typical electronic tools

As system designer, you will need **small tools** to assemble microcontroller projects and you will also need a **source of electricity** to check that they work correctly. You must work **safely** with these tools.

Power supplies and sources

All microcontrollers and most input/output devices are designed to work from **low voltage** and **low currents**. As the system designer, you must make sure your system is not connected to voltages or currents that are too high, as this could damage the components. Any power supplies used should conform to the relevant **safety standards**.

IF THE MICROCONTOLLER SYSTEM INTERFACES WITH HIGH VOLTAGE DEVICES, YOU MUST TAKE APPROPRIATE PRECAUTIONS.

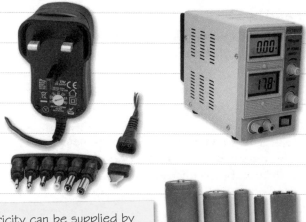

Electricity can be supplied by power supplies or batteries.

Small hand tools

The tools shown here are all designed to work on **low voltage systems** and must **not** be used where high voltages are present. You should use the tools only for the function they are designed for. This will minimise any health and safety risks.

You can use tools such as these to assemble the microcontroller and input / output devices.

Electrostatic discharge

(ESD) precautions

Some of the components used in microcontroller systems can be damaged by **static electricity**.

If you are using these components, take steps to remove any static charge. For example, the equipment shown opposite allows static charge to be conducted safely to earth.

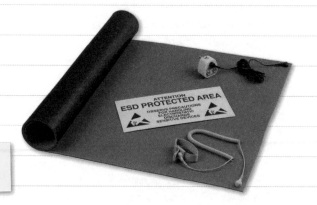

Electrostatic discharge equipment is used to prevent damage to components.

Now try this

Describe the hazards that these small hand tools could present and how to prevent any harm to the user.

Assembling and operating a microcontroller system

This page revises methods of **prototyping circuits**.

Breadboards

A breadboard provides a **quick method** of assembling prototype circuits to test their performance.

- Connections are made between components by pushing wires into the holes on the breadboard.
- Take care when using some components on a breadboard as they could be damaged by electrostatic discharge.

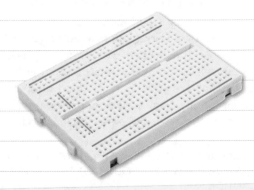

 Links See page 191 to revise electrostatic discharge (ESD) precautions.

The **breadboard** can be used to connect a LED to an Arduino™ Uno board. The rows in green show four separate rows of holes used to connect the circuit to a power source. The blue section shows how columns can be joined together.

Input/output modules

Input/output modules are available for some development platforms.

- A module provides a **ready-made assembly** of the components required to connect the hardware with the microcontroller.
- The process of connecting modules varies from platform to platform but all are very simple. This ease of assembly allows you to focus on writing the **control software**.
- Using modules can be more expensive than using discrete components.

An Arduino™ **radio module** ready to be connected. All the user has to do is align the connecting pins and push the module into place.

Prototyping boards

Preassembled prototyping boards are available for some development platforms.

- These provide the system designer with a range of **input** and **output** devices.
- If the actual device the system requires is not present on the prototyping board, you can use a device that functions in the same manner to check that the software will work.

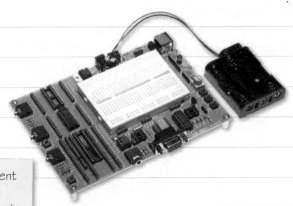

On this **PICAXE® experimenter board** there are four different output LEDs. These could be used to simulate any four devices that simply need to be turned on to make them work.

Now try this

If you were to design one of these experimenter boards, name three input and three output devices you would want to be built onto the board.

PICAXE® and Logicator:
program files and error checking

On this page you will revise how to start to **program** a PICAXE® microcontroller using BASIC software. See page 194 to revise **compiling** and **debugging**.

Connecting to microcontroller hardware

When Logicator opens:

- First check that the software can **connect** to the hardware. This is controlled by the **settings** window shown here.

- Choose the **type** of PICAXE® microcontroller you will use and where the programming lead will **connect** to the computer.

- If there are any **problems** with the settings the drop-down menus will be grey, showing where the problem is.

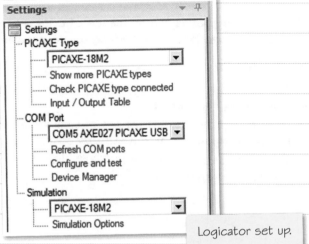

Logicator set up.

Creating and managing program files

Once the hardware is connected you can start the **software writing** part of the project.

1. Decide whether to write in **BASIC** or using a **flowchart**. Flowcharts have a restricted set of commands and can be complicated for large programs.

2. The development of a microcontroller system will be an **iterative process** with a number of different versions of the program being produced. So it is important to use a logical and easy-to-follow **system** of naming and saving files. An iterative process is one that **refines ideas** to develop the **best solution**.

3. You will then **type in** the **program**.

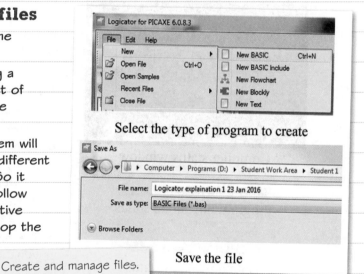

Select the type of program to create

Save the file

Create and manage files.

Syntax/error checking

Having written the program, carry out **syntax checking**. In Logicator the F4 key will check the syntax of the program.

- If any **errors are present** a message will be displayed indicating the type, and location, of the error.

- Once all the errors have been corrected a **successful** message is displayed.

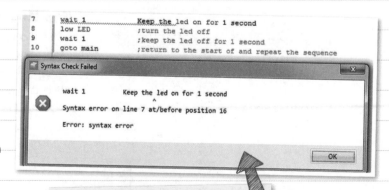

Syntax checking ensures that the structure of the program meets the 'rules' of the programming language.

Now try this

Describe 'syntax error' and explain how syntax errors are detected.

PICAXE® and Logicator: simulation, compiling and debugging

This page **completes** the sequence of operations needed to program a PICAXE® microcontroller using BASIC software. See page 195 to revise **creating program files** and **error checking**.

Simulation

When the syntax is correct you can check that the program performs as required. This can be achieved by **simulating the circuit**.

- Logicator provides two modes of simulation. The program can either run from start to finish, or each line of program can be run step by step. This is the method shown here.
- The graphic display of the PICAXE® chip also provides an indication of what is happening on each of the microcontroller pins.

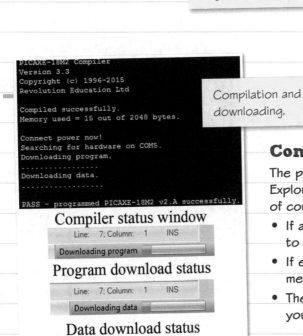

Logicator simulation.

Compilation and downloading.

```
PICAXE-18M2 Compiler
Version 3.3
Copyright (c) 1996-2015
Revolution Education Ltd

Compiled successfully.
Memory used = 15 out of 2048 bytes.

Connect power now!
Searching for hardware on COM5.
Downloading program.
. . . . . . . . . . . . . . . .
Downloading data.
. . . . . . . . . . . . . . . .
PASS - programmed PICAXE-18M2 v2.A successfully.
```

Compiler status window

Line: 7; Column: 1   INS
Downloading program

Program download status

Line: 7; Column: 1   INS
Downloading data

Data download status

Compiling, downloading and live testing

The program can now be **compiled**. The Workspace Explorer window provides information about the progress of compilation.

- If any **problems** occur the message displayed allows you to identify the corrections needed.
- If everything is **correct** the window will display the message PASS.
- The program is then sent to the microcontroller, where you can check the **real-life performance** of the system.

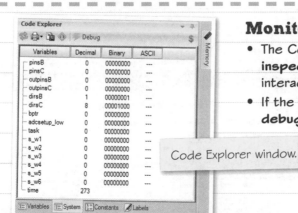

Code Explorer window.

Monitoring/debugging

- The Code Explorer window of Logicator allows you to **inspect** a range of information about how the program interacts with the microcontroller.
- If the system does not perform as required this will help **debug** the software to ensure it functions correctly.

Now try this

Explain the meaning of 'software compiling'.

Consider the role of the designer and the operation performed by the microcontroller.

194

Microchip, PICkit 3 and MPLAB®: program files and error checking

On this page you will revise how to start to **program Microchip, PICkit 3 and MPLAB®**. See page 196 to revise **simulating** and **debugging**.

Starting a Microchip project

As there is a large range of microprocessors available from Microchip, you have several **choices** to make at the start of a project:

- the **type of project** to be made
- the specific **microcontroller chip** that will be programmed
- if specific **hardware** is available to assist with **debugging**
- the **type** of hardware that will be used to **program** the microcontroller
- the specific **compiler** that will be used to **produce the code** that will be loaded onto the microcontroller
- the **name** for the project and the where to **store** the files produced.

Managing program files

The next stage of the process is to add the header file.

- **The header file** provides the program with the details needed to interface with the microcontroller.
- **The source file** is the program the system designer writes.

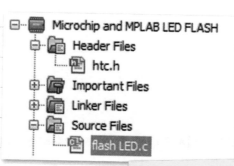

Header and source file menu.

Creating program files, checking syntax, and connecting to hardware

As you type in the program code, hints, tips and possible errors are indicated by symbols replacing line numbers. This interactive help allows you to **correct the code** as you enter it, rather than waiting until the end. The **project dashboard**, shown on the left, tells you that the relevant **hardware is available** for sending the program to. It also provides **details of the choices** made during the opening options.

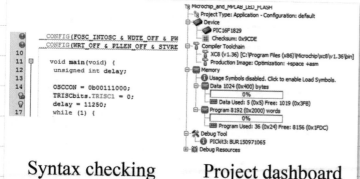

Syntax checking and project dashboard.

Now try this

Explain why it is an advantage for the system designer that code is checked as it is entered.

Microchip, PICkit 3 and MPLAB®: simulation and debugging

This page **continues** the sequence of operations needed to program a Microchip microcontroller using the MPLAB® X. See page 195 to revise **creating program files** and **error checking**.

Simulation

When you have finished writing the program:

- you can **test** it using the **project properties** menu
- select the **simulator option** rather than the hardware programmer chosen during the initial set-up options.

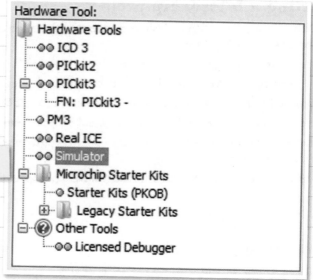

Simulator

Monitoring/debugging, downloading and live testing

If simulation confirms the system performs as expected, the program is made and transferred to the **connected microcontroller**. A feature of the **Microchip PICkit 3** is that the program can be **stored** onto the programmer. This then allows the PICkit 3 programmer to **transfer** the program to **multiple microcontrollers** at the touch of a button on the PICkit 3 hardware.

```
Connecting to MPLAB PICkit 3...

Currently loaded firmware on PICkit 3
Firmware Suite Version.....01.40.13
Firmware type..............Enhanced Midrange

Programmer to target power is enabled - VDD = 5.000000 volts.
Target device PIC16F1829 found.
Device ID Revision = 4

The following memory area(s) will be programmed:
program memory: start address = 0x0, end address = 0x7ff
configuration memory

Device Erased...

Programming...
Programming/Verify complete
```

Microcontroller programmed.

Now try this

Explain why it is an advantage to be able to transfer programs directly from the PICkit 3 hardware.

GENIE®: program files

On this page you will revise how to start to **program** a GENIE® microcontroller using **programming editor software**. See page 198 to revise **simulation** and page 199 to revise **compiling** and **debugging**.

Connecting to microcontroller hardware

- If the computer system is **not connected** to the GENIE® microcontroller, a troubleshooting routine can be run. This will **automatically** set up the computer system for correct operation.

- When the computer system is **connected**, the menu tells you the **type** of GENIE® microcontroller available and offers a number of **options** about what to do next.

When the GENIE® programming editor loads, one of these windows will be shown in the program tab.

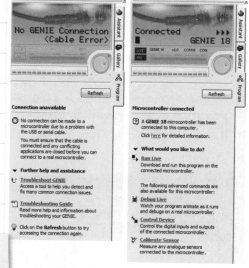

Genie not connected

Genie connected

Creating and managing program files

Once the hardware is **connected** you can start writing the software.

- **Select** the correct GENIE® microcontroller to produce a flowchart for.
- **Save** the flowchart, remembering to use logical systems for file-naming and saving.
- **Produce** the flowchart by dragging symbols from the gallery onto the drawing area. **Customise** these symbols to create the required action by right-clicking on them.

Produce the file.

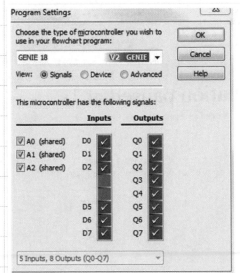

Select the type of microcontroller to write a program for

Save the file

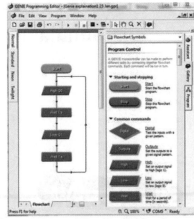

Draw the flowchart

Now try this

Explain one advantage and one disadvantage of using flowcharts to program a microcontroller.

GENIE®: simulation

On this page you will revise how to **simulate** a program on a GENIE® microcontroller using programming editor software. See page 199 to revise **compiling** and **debugging**.

Simulation

When you have drawn a flowchart, check that it works as expected.

- With a GENIE® flowchart there is no need to check syntax errors, but check that the program is **appropriate** for the **connected microcontroller** using the **program menu**.
- The simulation **displays the state** of each of the **inputs / outputs** for each step of the flowchart.
- As the program runs, each **symbol** on the flowchart becomes **highlighted**. You can **pause the program** at any stage to consider what is happening with the flowchart and microcontroller.
- If the program does not function as expected, **edit the flowchart** to correct the problem.

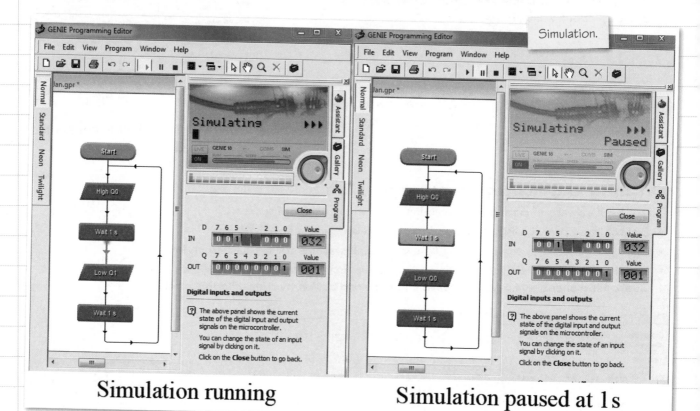

Simulation.

Simulation running   Simulation paused at 1 s

Explain why a program should be simulated before being downloaded to the microcontroller.

GENIE®: compiling and debugging

This page **completes** the sequence of operations needed to program a GENIE® microcontroller.
See page 197 to revise creating **program files** and page 198 to revise **simulation**.

Compiling, downloading and live testing

When simulation has been completed:

* the program can be **sent** to the microcontroller
* as system designer, you can **inspect** the correct operation of the physical system.

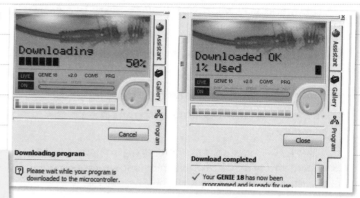

> The program window shows the progress of the download and confirms completion.

Debugging

The **Debug Live** option allows you to inspect the current state of inputs and outputs of the microcontroller.

* As the program runs, the **symbol** on the flowchart becomes highlighted in sequence. The **inputs** and **outputs** associated with the symbol are shown as **red 0** (off) or **green 1** (on).
* If the microcontroller system does not operate as expected, this will help you **identify** where any problems are so that you can correct them.

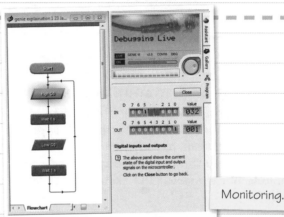

Monitoring.

Highlight shows　State of output
current step　　at current step

Monitoring

If required:

* you can **expand** the range of inputs and outputs
* you can **examine** the **values** of analogue sensors and variables, as shown opposite
* you have a useful facility in the ability to **slow down** the speed that the program runs at. The controls for this feature are shown below.

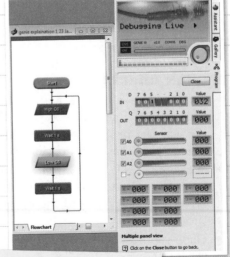

Monitoring of multiple factors.

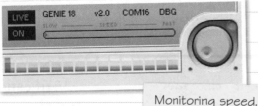

Monitoring speed.

Arduino™ Uno: program files and error checking

On this page you will revise how to start **programming** an Arduino™ Uno. See page 201 to revise **simulation** and **testing**.

Connecting to the microcontroller

The **tools** menu is used to:

- select the **type of board** that is being used
- see how the microcontroller **connects** to the computer
- select the type of **external programmer** to be used; an external programmer allows more of the program space on the microcontroller to be used.

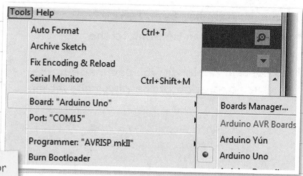

When the Arduino™ programming editor is started this screen is displayed.

Creating and managing program files

Arduino™ programs are called **sketches**.
A template for a sketch automatically appears, with the essential sections already defined. If the program needs to include any libraries these can be selected from the **Include library** menu. The sketch can then be **saved**.

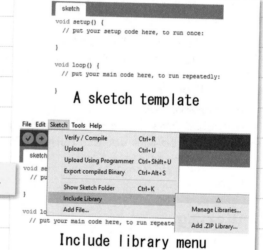

A sketch template

Sketch template and Include library menu.

Include library menu

Syntax checking

- When you have finished writing the sketch it has to be **verified** to ensure the structure is valid. If there are any problems with the way the program has been written the software will show where, and what, the problem is.

- In this example, the system designer has incorrectly written 'high' as opposed to the correct 'HIGH'. The line with the error is highlighted and the **error message indicates** what the problem is.

- Once the sketch is **correct** the '**done compiling**' message will be shown.

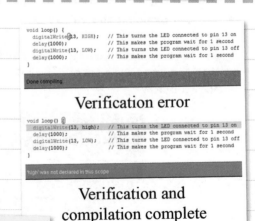

Verification error

Verification and compilation complete

Syntax checking uses the Verify / Compile option.

Now try this

Explain what a library is in relation to the Arduino™ platform.

Arduino™ Uno: simulation and testing

This page **completes** the sequence of operations needed to **program** an Arduino™ Uno. See page 200 to revise creating **program files** and **error checking**.

Simulation

There is **no simulation facility** in the Arduino™ programming environment.

- The **sketch can be simulated** using **other** software or online.
- The **free website** Autodesk 123D Circuits (https://123d.circuits.io) allows you to construct a model of your input / output circuits and use your own code to **simulate** the operation of the complete system.

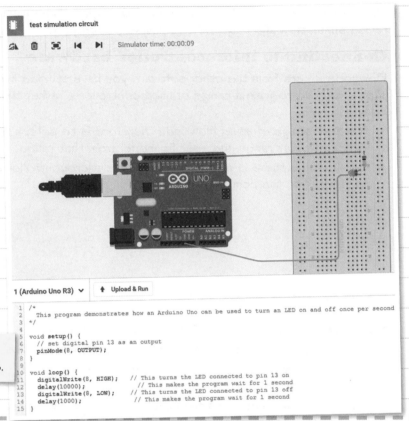

Online Arduino™ **simulation** using Autodesk 123D Circuits.

Compiling, downloading and live testing

Once the simulation confirms that the program performs as expected, the Arduino™ microcontroller can be **programmed**.

- When you click the upload icon, the program is **verified**, **compiled** and **sent** to the microcontroller. If this process is successful the image on the right will be shown.
- Note that, depending on the simulation method, it may be quicker to **test the real circuit** than run the simulation. To test the real circuit, all the hardware must be available.

Compile and upload complete.

Now try this

Explain how Arduino™ circuits and programs can be simulated.

Flowcode and E-Block: creating and managing program files

On this page you will revise how to use Flowcode to **program** a PIC® microcontroller. See page 203 to revise **simulation** and page 204 to revise **debugging**.

Connecting to microcontroller hardware

Flowcode differs from the other software you have studied because it allows you to **program a range** of microcontrollers, rather than a specific manufacturer's.

- **The first stage** of using Flowcode therefore is to **select** the target microcontroller range and specific model from that range.
- At this stage there is **no need** to have the microcontroller hardware connected to the computer.

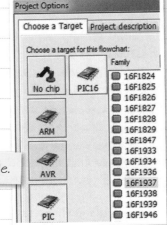

Choose a target in Flowcode.

Creating and managing program files

At the start of the project, select the **input** and **output** devices.

- This adds a **physical model** of the device that will be interactive during a simulation.
- It will also make sure the **appropriate code** is available to control the device.
- As well as conventional input/output devices, Flowcode provides models for a wide range of **other devices**; for example, Bluetooth and RFID.
- The program is **saved** through the **file menu**.

Flowchart production.

Select input devices

Select output devices

Creating program files

Produce the flowchart by **dragging** the **flowchart symbols** from the **menu** onto the **drawing area**. Flowcode works differently from many flowcharting methods in that you do not need to draw the connections between the symbols. As you place the symbol in position, the software **automatically adds the connections**. Right-clicking on a symbol allows its properties to be **modified**.

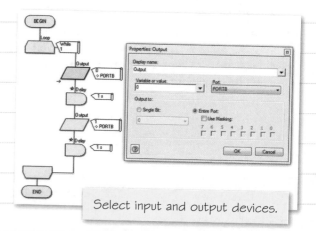

Select input and output devices.

Now try this

Explain why the ability of Flowcode to program multiple types of microcontroller is an advantage for the system designer.

Flowcode and E-Block: simulation

This page revises the **simulation** needed to program a PIC® microcontroller using Flowcode. See page 202 to revise **creating** and **managing** program files and page 204 for **debugging**.

Simulation

Flowcode offers a **different approach** to simulation compared with other software you have studied.

- It provides a **choice** of 2D or 3D representations of the **target output device** instead of showing the state of the microcontroller output pins.
- First, **select the type of device** to simulate. You then **configure** the device (e.g. by choosing the colour of the LED) and **specify** which output of the microcontroller controls the device.
- When the **simulation** is run, the model of the object will **behave in a similar way** to the object it represents. For example, a LED will become brighter when it is turned on by the microcontroller.

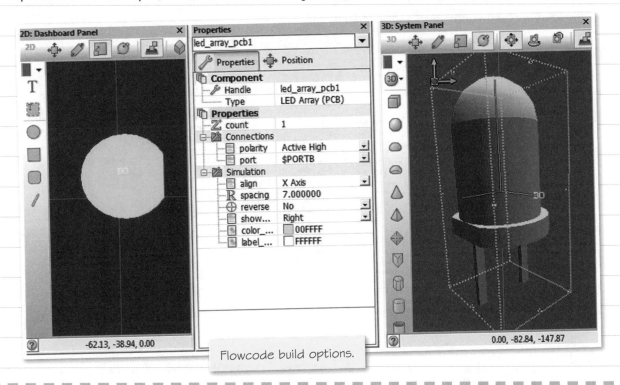

Flowcode build options.

Compiling, downloading and live testing

Flowcode sends the program to the microcontroller in **three stages**.

- All three stages can be completed in a **single step**, by selecting 'Compile to chip'.
- This **converts** the visual image into **C**, then into **assembly**, then to **hexadecimal** and, finally, sends this code to the microcontroller.
- The alternative is to **compile the code** to one of the **intermediate stages**. This allows the expert system designer to **refine** the code to add features or refine the way the program operates.

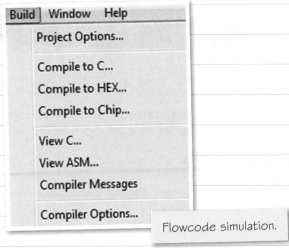

Flowcode simulation.

Now try this

Explain one advantage and one disadvantage of Flowcode's methods of simulating a system.

Flowcode and E-Block: simulation and debugging

This page **completes** the sequence of operations needed to program a PIC® microcontroller using Flowcode. See page 202 to revise **creating** and **managing** programs and page 203 to revise **simulation**.

Monitoring/debugging

Flowcode offers several extra features compared to other platforms.

- Some of the **E-blocks** are available with **Ghost** technology built into them to **analyse** system performance.
- Ghost technology reads the **actual physical values** of the pins on the microcontroller and can **display** them on a simulated oscilloscope on the user's computer.
- This allows **accurate analysis** of complex tasks.

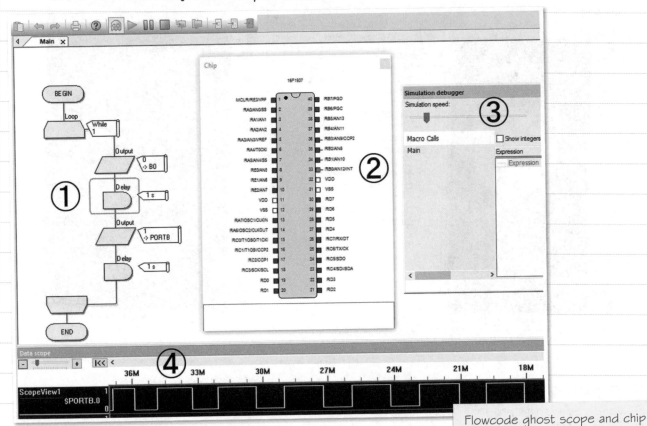

Flowcode ghost scope and chip

In the image above, ① is the flowchart that is being run on the microcontroller.

② shows the status of the individual pins on the microcontroller. In this instance, the red square indicates that pin 33 is high.

The speed at which the flowchart runs can be altered by dragging the slider, shown at ③.

The oscilloscope display that shows how the pin's status changes is shown at ④.

Now try this

Describe the function of an oscilloscope.

Coding practice and efficient code authoring

Comments

Comments provide information in an easy-to-read format that records and explains key features of the program.

- **Introductory comments** provide an overview of the complete program.
- **In-line comments** relate to specific sections of the program.
- **Some introductory** comments have to follow a **style guide** set by the end users of the software.

Style guides

For example, NASA requires that the introductory comments include information about the file name, a description of the purpose of the file, a list of the files used by the program and the development history of the program.

Chip set-up and pin modes

For some microcontroller platforms, such as Microchip, the microcontroller must be specifically configured to function the way the system designer wants it to by setting the **configuration bits**. These are set only when the microcontroller is programmed and remain in the same state unless the microcontroller is reprogrammed. Configuration bits control such aspects as how fast the microcontroller operates.

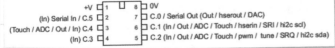

Address	Name	Value	Field	Option	Category	Setting
8007	CONFIG1	FFFF	FOSC	ECH	Oscillator Selection	ECH, External Clock, High Power Mode (4-32 MHz): device clock supplied to CLK...
			WDTE	ON	Watchdog Timer Enable	WDT enabled
			PWRTE	OFF	Power-up Timer Enable	PWRT disabled
			MCLRE	ON	MCLR Pin Function Select	MCLR/VPP pin function is MCLR

MPLAB® X and PIC16F1829 Configuration Bit Menu.

Pin modes

The pins on a microcontroller are the physical connections that allow it to **interact** with external components. They can be **configured** to perform different functions at the start of the program by the system designer.

```
            PICAXE-08M2
              +V ☐ 1    8 ☐ 0V
  (In) Serial In / C.5 ☐ 2    7 ☐ C.0 / Serial Out (Out / hserout / DAC)
(Touch / ADC / Out / In) C.4 ☐ 3    6 ☐ C.1 (In / Out / ADC / Touch / hserin / SRI / hi2c scl)
        (In) C.3 ☐ 4    5 ☐ C.2 (In / Out / ADC / Touch / pwm / tune / SRQ / hi2c sda)
```

The pins for one microcontroller. Pin number 5, labelled C.2, can be set up to act as a digital input or output, an analogue to digital convertor, a touch sensor, a PWM output, a sound output, a latch or a connection to an I²C bus.

Libraries – declarations

Libraries are a way of extending the functionality of a programming language.

- The C language has a range of **standard libraries**, defined by organisations such as ISO and ANSI.
- Additional libraries that provide specific functionality are also available. To make use of these libraries they must be added to the program. For example, the line **#include <stdio.h>** allows the system designer to make use of standard input / output functions.
- A **declaration** is the method by which the system designer assigns a name, or identifier, to a variable or function. A declaration has the 'specify the data' type. For example, 'int wholeNumber' declares that the identifier 'wholeNumber' is of the integer type.

Code syntax

Syntax refers to the **set of rules** that programs must follow in order to be valid.

- For **text-based** languages, such as BASIC and C, the syntax rules relate to the order of the words and characters the system designer enters.
- For **visual-based** languages, such as flowcharts, syntax rules relate to the ways the symbols are connected.

Code organisation and structure

As the programs written by the system designer become more complex, it is important to use a **systematic** and **logical** programming style. How you achieve this style will vary depending on the programming language. A key consideration is that the operation program should be **transparent** to the writer, and to potential readers, of the program.

Now try this

Write a set of introductory comments that provide an overview of a program that you have written previously.

Coding constructs: inputs and outputs (BASIC and C) 1

The table below and the one on the following page illustrate how to achieve a range of **input** and **output** functions using BASIC and C.

Programming construct	PICAXE® BASIC	Arduino™ C
Digital – port level read/write	To configure a pin on a specified port as an input, or output, a command like this is used: `let dirsA = %00001111` This sets pins 7–4 as inputs and pins 3–0 as outputs on port A.	The DDR function configures a port so that the pins of the port are read or write. `DDRB = B111000` This sets port B such that pins 13 to 11 are write and pins 10 to 8 are read.

Digital – bit level read/write	To turn a pin on, or off, a command like this is used.
	`Let pinsA = %00001111`
	This turns pins 7–4 off, and pins 3–0 on.
		The pinMode function configures individual pins to be either inputs or outputs.
		`pinMode (13, OUTPUT)`
		The function above sets pin 13 as an output.
		The digitalWrite function is used to turns outputs on or off.
		`digitalWrite (13, HIGH)`
		The function above turns pin 13 on.

Analogue read 8-bit resolution	To read the value at the analogue to digital convertor, a command like this is used. `readadc C1, b1` This reads the value at pin C1 and stores it in the variable b1.	For the Arduino™ Uno microcontroller, all analogue to digital conversions are performed with 10-bit accuracy. The function used to perform this is analogRead. In order to use the data obtained by the ADC function, the value must be assigned to a variable. `adcvalue = analogRead(3)`
Analogue read 10-bit resolution	This command reads the ADC with a resolution of 10 bits. `readadc10 C1, w1` This reads the analogue value at pin C1 and stores it in the wordvariable w1.	The function above converts the analogue signal on pin 3 to a digital value. It then assigns this value to the variable named adcvalue.
Calibration	This command ensures that the value from the ADC is accurate. `calibadc b1` This will provide an ADC reading that can be compared to a known fixed voltage.	There is no direct equivalent of this command in Arduino™ C. In order to calibrate a sensor, the maximum and minimum values are determined using the ADC. The map function can then be used to ensure that these values are assigned to Start and End of the ADC range.
Tone and sound generation	To cause a tone to be generated, a command like this is used. `sound B7, (50,50)` This generates on pin B7 a tone of 50 Hz for a duration of 500 ms.	To cause a tone to be generated, a command like this is used. `tone(7, 440, 500)` This generates on pin 7 a tone of 440 Hz for a duration of 500 ms.

Now try this

Consider the programming construct used for bit level read and write. Describe one situation where the format of the PICAXE® command would be easier to use and one where the format of the Arduino™ C command would be easier to use.

Coding constructs: inputs and outputs (BASIC and C) 2

The table below continues from page 206 to illustrate how to achieve a range of input and output functions using BASIC and C.

Programming construct	PICAXE® BASIC	Arduino™ C

Pulse and pulse width modulation (PWM)	A command like this configures a pin to provide a PWM output.
	`pwmout C2, 150, 75`
	This command outputs a PWM signal on pin C2, with a period of 150 ms and a duty cycle of 75 Hz. There are wizards available within the PICAXE® editor to calculate appropriate values for the period and duty cycle.
	`pwmduty C2, 75`
	This command alters the duty cycle.
		Within the Arduino™ platform PWM is referred to as analogWrite. This function generates a square wave on the designated output pin.
		`analogWrite(3, 128)`
		This function will generate a PWM signal on pin 3 with a duty cycle of 50%. Duty values range from 0 (always off) to 255 (always on).

Serial communication	A command like this causes a serial signal to be transmitted.
	`serout 7, N2400, (100)`
	This command transmits the signal from pin 7, at a speed of 2400 baud and sends the data 100.
	To receive a serial signal the command below is used.
	`serin 7, N2400, b1`
	This command receives the signal on pin 7, at a speed of 2400 baud and stores the data in the variable b1.
		In order to use serial communication on an Arduino™ microcontroller, it must first be configured using the command.
		`Serial.begin (speed)`
		The function below will transmit the word Hello via the serial link.
		`Serial.write("Hello");`
		The function below will receive data via the serial link and store it in the variable incomingByte.
		`incomingByte = Serial.read();`

I^2C communication	The PICAXE® microcontroller needs to be configured as either a slave, or master. This is achieved through the commands below.
	`hi2csetup i2cslave %10100000`
	`hi2csetup i2cmaster %1010xxxx`
	`i2cfast i2cbyte`
	To read data from an I^2C device, the command below is used.
	`hi2cin`
	To send data to an I^2C device, the command below is used.
	`hi2cout`
		This function is available through the wire library.
		The function below connects the microcontroller I^2C bus.
		`Wire.begin()`
		After a range of other parameters are set, the function below will send the number 100.
		`Wire.write (100)`
		To receive the number, the function below would be used.
		`int number = Wire.read()`

Now try this

Investigate the wizard available in the PICAXE® editor to calculate pulse width modulation values.

Coding constructs: inputs and outputs (GENIE® flowchart) 1

This page revises the range of input and output functions using GENIE®, continuing to page 209.

Digital – port level read/write

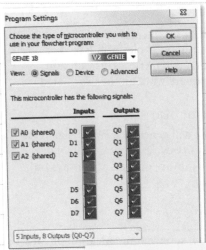

Input and output ports are fixed and determined by the type of microcontroller being programmed.

Digital – bit level read/write

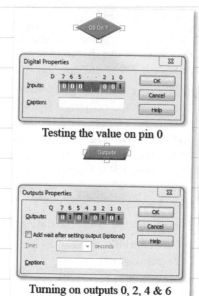

Testing the value on pin 0

Turning on outputs 0, 2, 4 & 6

GENIE® input/output configuration: the digital symbol allows input pins to be read, and the output symbols turn pins on or off.

Analogue read 8-bit resolution

In A, A0

The input symbols allow either an analogue or digital signal to be read and assigned to a variable.

Calibration

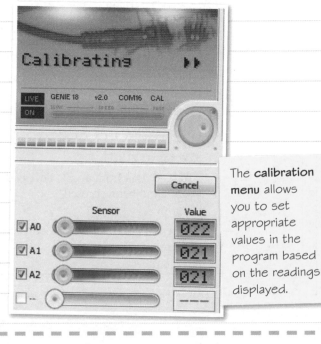

The calibration menu allows you to set appropriate values in the program based on the readings displayed.

Now try this

Whether a port is an input or output is predetermined on the GENIE® microcontroller. Explain one advantage and one disadvantage of this.

Coding constructs: inputs and outputs (GENIE® flowchart) 2

This page revises the range of input and output functions using GENIE®, continuing from page 208.

Tone and sound generation

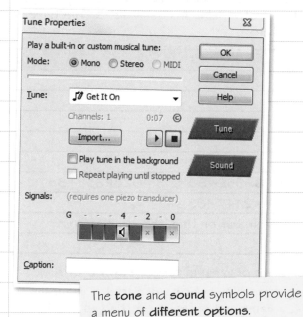

The **tone** and **sound** symbols provide a menu of **different options**.

Pulse and pulse width modulation (PWM)

The **pulse symbol** allows you to choose between a **single pulse**, a **pattern of pulses** or **PWM**. Options to configure each of these are presented in the drop-down menus.

Serial communication

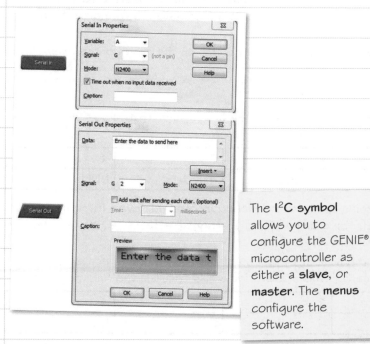

The I²C symbol allows you to configure the GENIE® microcontroller as either a **slave**, or **master**. The **menus** configure the software.

Communication I²C

The **serial in** and **serial out** symbols present a **range of options** to control their function.

Now try this

Practise using GENIE® to construct a flowchart that sends a single pulse of 100 ms duration.

Coding constructs: logic and arithmetic variables and arrays

Variables

A variable is a way of **storing values** within a program. Variables have a wide range of uses; for example, they can be used to hold data from sensors, or the results of calculations. Depending on the microcontroller platform being used, a number of different types of variable will be available. The table below shows the types of variable available for the **Arduino™ Uno**.

Variable	
Boolean	A Boolean holds one of two values: true or false.
Character	A single alphabetic character, such as A, B or C
Byte	An integer number in the range 0 to 255
Integer	An integer number in the range –32768 to 32767
Word	An integer number in the range 0 to 65,535
Float	A number with a decimal point in the range 3.4028235E+38 to 3.4028235E+38
Long	An integer number in the range –2147483648 to 2147483647
Double	Generally, a floating number with twice the precision, but on Uno, the same as float
String	A series of characters, such as 'ABC'

- **GENIE®** and **PICAXE®** microcontrollers have a predefined range of variable names; e.g. B0 or A.
- Before variables can be used in **C-based** programing languages they must be declared. This means defining the type of **data** the variable will hold, allocating a **name** and optionally setting an initial **value**. For example, within an Arduino™ program the following line would declare that 'myvariable' will hold integer-type data and is set to a value of 3 initially:

  ```
  int my_variable = 3
  ```

- Should it be needed, it may be possible to convert data from one type to another. For example, a **float** number could be converted to an **integer**. This would reduce the memory needed to store the data and would allow the program to run faster. Another common type of data conversion is to change **integers** to **ASCII** characters and vice versa.
- Data can be manipulated by a range of commands, the number dependent on the microcontroller platform used.

Arrays

An array is a programming **construct** that is used to store **variables** in a manner that allows them to be **easily accessed**. A graphical representation of an array is shown below.

Index number	0	1	2	3	4
Data	1st data	2nd data	3rd data	4th data	NULL

- By referring to the arrays, index number-specific data can be placed into or retrieved from, memory. In order to use an array, it must be declared. For example, int my_array[5] = {2, 4, 6, 8}; will declare an array named my_array to hold the four integers 2, 4, 6, 8.
- The size of the array must be one larger than the number of pieces of data to be held, because the end of the array must be indicated by the null character. This declaration would result in the following array.

Index number	0	1	2	3	4
Data	2	4	6	8	NULL

- Data can be read from this array by using the command my_array [2] which, in this example, would retrieve the third data element, the number 6.

> ### Now try this
>
> The methods shown above for arrays are for C-based programming languages. Research how arrays, or their equivalents, can be used in GENIE® or PICAXE® microcontrollers.

Coding constructs: logic and arithmetic

Comparative and **Boolean operators** are often used in combination with microcontroller systems to determine the way in which the system operates.

Comparative operators

You will often require software to perform a different **set of instructions** depending on the status of particular **variables**. Comparative operators allow the program to **compare variables** to each other or to constants, and then move to different parts of the program, depending on the outcome of the comparison.

Comparative operator	Symbol	Meaning
equal	= =	x equal to y
not equal	! =	x not equal to y
less than	<	x less than y
more than	>	x more than y
less than or equal	< =	x less than or equal to y
more than or equal	> =	x more than or equal to y

Boolean operators

Boolean operators control the function of a program by **testing** the state of **inputs** or **variables**. The **answer** to the test determines how the program proceeds.

- The Boolean operators AND / OR combine two inputs into a single output.
- The NOT Boolean operator reverses the state of an input.
- The tables below are called **truth tables** and show what the output state will be, depending on the results of the tests performed on the inputs.

AND

Input 1	Input 2	Output
False	False	False
True	False	False
False	True	False
True	True	True

OR

Input 1	Input 2	Output
False	False	False
True	False	True
False	True	True
True	True	True

NOT

Input 1	Output
True	False
False	True

Boolean operators are often **combined** with comparative operators in a microcontroller program.

For example, for the four inputs A, B, C and D with values of 1, 2, 3 and 4, respectively, a typical Boolean test might be:

If A < B AND C < D do X.

In this case, since A (1) is less than B (2) AND C (3) is less than D (4) the output is True and X would happen.

Logic using analogue input conditions

For microcontroller systems that make use of analogue inputs, absolute values of the inputs may be hard to predict. The comparative operators > and < allow a **wide range** of values to be used.

Logic using digital input conditions

When digital inputs are used it is much easier to predict the input values exactly. In this case, the more **specific** comparative operators such as = = can be used.

Now try this

Produce AND, OR and NOT truth tables for a range of input conditions of your choice.

Coding constructs: program flow and control 1

You can control the operation of a program using **subroutines**, interrupts and delays. Subroutines are revised on this page, and interrupts and delays are revised on page 213.

Subroutines

Choose names for subroutines that reflect the **tasks** the subroutine **performs**. In the examples below, 1 is where the subroutines are declared, and 2 is where the subroutines are called or used. The program below will play either a 'rock' or 'pop' sound, depending on the state of an input.

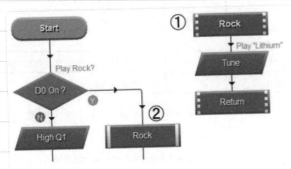

Flowchart subroutines

```
        if pinC.7=1 then gosub Rock
                                  ②
        goto main
Rock: ①
        sound B.0, ( 3, 2 )
        return
```

Basic Subroutines

> Subroutines are a way of **grouping together** a series of **commands** that can be **reused** by different parts of a program.

C functions

```
int inPin = 7;              // Switch is connected to digital pin 7
int switchState = 0;        // Variable to store the state of the switch

① void Rock(){              // This creates the function called Rock
     tone(7, 200, 4);}      // Play a note of frequency 200

② void Pop(){               // This creates the function called Pop
     tone (7, 400, 4);}     // Play a note of frequency 200

   void setup()
   {
       pinMode(inPin, INPUT);   // Sets the digital pin 7 as input
   }

   void loop(){
   {
     switchState = digitalRead(inPin);  // Read the switch
③    if (switchState = 1 ) Rock;        // If the switch is on call the Rock function

④    if (switchState= 0 ) Pop;          // If the switch is off call the Pop function
   }
   }
```

> Example C functions. In C, functions perform the **same role** as subroutines in a flowchart or BASIC. **1** and **2** are where the functions are **declared**. **3** and **4** are where the functions are **called** or used.

Now try this

In the C program above, the functions are defined as being of the data type 'void'. Research and explain what 'void' indicates.

Coding constructs: program flow and control 2

You can control the operation of a program using subroutines, **interrupts** and **delays**. Subroutines are revised on page 212 and interrupts and delays are revised on this page.

Calling libraries

- With the **PICAXE®** and **GENIE®** microcontroller platforms the range of tasks that the software can perform is **fixed**.

- With platforms that **use C** as the programming language there is a standard **range** of functions the software can perform. These are parts of the **standard C library**, the contents of which are specified by organisations such as ANSI and ISO.

- As well as the standard libraries, the system designer can add, or call, **extra ones** to provide extra functionality. For example, the **Arduino™** library called NewPing makes it easier to work with ultrasonic sensors, such as the SRO4.

- In order to use the functions provided by any additional libraries, you must **specify** them at the start of the program using this syntax: #include <libraryName>

Delays and timing

The speed at which a microcontroller functions can sometimes be too fast. For example, a series of numbers could be displayed faster than a human can perceive an individual number.

- **Delays** allow the microcontroller to complete tasks in a suitable time frame. Depending on the platform being used, delays can range from fractions of a second to hours.

- For some applications you will need your program to complete **time critical** tasks. A **timer** is a hardware feature that allows an accurate measurement of the time elapsed between two events.

Interrupts

- An **interrupt** is a method of interrupting the normal execution of a microcontroller program. It can be triggered by a variety of events, such as a user pushing a button.

- An interrupt could be used within a program that makes use of long delays.

- Normally during the delay period the microcontroller could not react to any inputs.

- By using an interrupt you can create an event that will terminate the delay and allow **another task** to be performed.

- Dependent on the microcontroller platform, interrupts can be implemented either through dedicated hardware **input pins** or **software**.

Application

The image shows a typical control panel for a microwave cooker. This control panel links to the microwave's microcontroller. This demonstrates the application of delays, timing and interrupts.

- **Delays** allow the user to set the time when they want the microwave to start cooking.

- **Timing** determines how long the microwave stays on for.

- **Interrupts** stop the normal operation of the microcontroller program if the door to the microwave is opened.

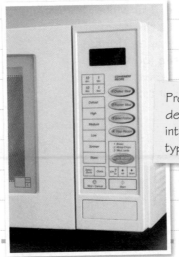

Programming for delays, timing and interrupts on a typical control panel.

Now try this

Explain how delays can be achieved on a platform of your choice.

Program flow: iteration

You can use **loops** to make a program **repeat** the same sequence of commands several times.

Flowchart loops

The flowcharts show three versions of the same program. The until loop will continue the sequence until the input changes. The while loop will continue the sequence unless the input changes. The repeat loop will continue the sequence a set number of times.

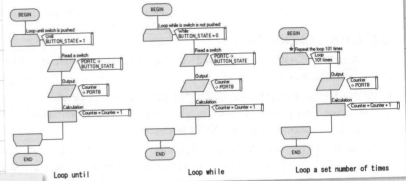

Loops in flowcharts.

BASIC loops

The code shows three versions of the same program written in **BASIC**. The until loop will continue until the input changes. The while loop will continue unless the input changes. The for - next loop will repeat the sequence a set number of times.

```
do
      Let pinsB = b1        ;Set the output pins to the value of b1
      Inc b1                ;Increase b1 by 1
loop until pinC.7=1         ;Repeat the loop until pin C7 is high
                     Loop until
do
      Let pinsB = b1        ;Set the output pins to the value of b1
      Inc b1                ;Increase b1 by 1
loop while pinC.7=0         ;Repeat the loop while pin C7 is low
                     Loop while

For b1 = 1 to 100           ;This will repeat the loop 100 times
      Let pinsB = b1        ;Set the output pins to the value of b1
next b1                     ;Add 1 to b.1 and return to the for command
                     For - next
```

Loops in BASIC.

C loops

The code shows the sections of C programs that perform similar functions to those above. The **while** loop performs the test at the start of the loop. The **do while** loop performs the test at the end of the loop. The **for - next** loop will repeat the sequence a set number of times.

```
} (Loop while)
void loop() {
int ledValue = 0;                  //set up an int variable ledValue
      while (ledValue <101){       // While ledValue is < 101 loop
      PORTB = ledValue;            // set port b to ledValue
      ledValue++;                  // increase ledValue by 1
      delay(100);                  // wait so pattern can be seen
      }
}

do (Loop do while)
{
      PORTB = ledValue;            // set port b to ledValue
      ledValue++;                  // increase ledValue by 1
      delay(100);                  // wait to allow the pattern to be seen
}
      while (ledValue <100);       // While ledValue is less than 100 loop
}

} (For - next)
void loop() {
for (int ledValue=0;ledValue<101;ledValue++) //Repeat 100 times
      { PORTB = ledValue;          //Set port b to ledValue
        delay(100);                //wait so pattern can be seen
      }
}
```

Loops in C.

Now try this

Using a platform of your choice, explain the differences between *loop while* and *loop until*.

Control of program sequence: if else

In order for a microcontroller system to **interact** with its environment, it will need to perform different actions depending on **operating conditions**. This page revises the *if else* control.

Using a flowchart to perform *if else* control

The *if* function provides **two different paths** for a program to follow based on the result of the condition being tested.

- **If the answer to the decision is yes**, the blue sequence is performed.
- **If the answer is no**, the red sequence is performed. The *no* response is the equivalent of the *else* statement in other programming languages.

Flowchart *if* control.

Using BASIC to perform *if else* control

This program demonstrates the use of the *if else* commands in BASIC.

- Line 2 tests the **state** of an input and if the answer is true the program moves to line 9.
- Line 4 is the *else* command that determines what happens if the answer is false.

BASIC *if else* control.

```
1    main:
2    if pinC.0 = 1 then         ; Test if pinC.0 is high
3            gosub flash         ; Goto flash if C.0 is equal to 1
4    else                       ; Do the next line if pinC.0 is not 1
5            gosub stayon        ; If C.0 is not equal to 1 stayon
6    endif                      ; Stop testing pinC.0
7    goto main                  ; Return to the start
8
9    flash:                     ; The subroutine called flash
10   low B.2                    ; Turn of output B.2
11   high B.1                   ; Turn on output B.1
12   pause 500                  ; wait 0.5 seconds
13   low B.1                    ; switch off output B.1
14   return                     ; go back to the line gosub flash
15
16   stayon:                    ; The subroutine called stayon
17   high B.2                   ; Turn on output B.2
18   pause 500                  ; Wait 0.5 seconds;
19   return                     ; Go back to the line gosub stayon
```

Using C to perform *if else* control

This program demonstrates the use of the *if* function in C.

① is the *if* function testing the state of the input button.

② is the *else* function that turns on a LED if the button is not in the high state.

```
const int buttonPin = 3;            // the pin the pushbutton is connected to
const int ledPin_button_on = 6;     // the pin the on LED is connected to
const int ledPin_button_off = 7;    // the pin the off LED is connected to

int buttonState = 0;                // variable for reading pushbutton status

void setup() {
  pinMode(ledPin_button_on, OUTPUT);   // setup pin 6 as an output
  pinMode(ledPin_button_off, OUTPUT);  // setup pin 7 as an output
  pinMode(buttonPin, INPUT);           // setup pin 3 as an input
}

void loop() {
  buttonState = digitalRead(buttonPin);           // Read the state of the input
  if (buttonState == HIGH) {                      // If the pushbutton is on then
    digitalWrite(ledPin_button_on, HIGH);         // Turn on the LED on pin 6
  }
  else {                                          // If the push button is not on
    digitalWrite(ledPin_button_on, LOW);          // Turn off the LED on pin 6
    digitalWrite(ledPin_button_off,HIGH);         // Turn on the LED on pin 7
  }
}
```

Arduino™ C *if* control.

Now try this

Explain the use of the *if else* flow control functions on a platform of your choice.

Control of program sequence: switch 1

This page revises the *switch case* and *else if* commands. Arduino™ *switch* and BASIC *elseif* are revised on this page and BASIC *switch* is revised on page 217.

Using a flowchart to perform *switch case* control

The *switch* function is similar to the *if* function except that **more than one** condition, or case, can be tested.

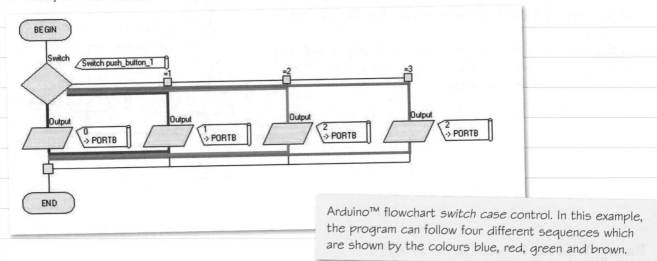

Arduino™ flowchart *switch case* control. In this example, the program can follow four different sequences which are shown by the colours blue, red, green and brown.

Using BASIC to perform *elseif* control

There is not a direct equivalent command to *switch* in BASIC. However, the *elseif* command allows multiple *if* commands to be used to achieve the same outcomes as the *switch* function.

```
1     main:
2     readadc C.1,b1                    ; Read the ADC and store the value in b1
3     if b1 <=10 then                   ; If the value of b1 <10 then
4           gosub small                 ; Goto the subroutine small
5     elseif b1>10 and b1<=20 then      ; If the value of b1 is >10 and =<20 then
6     gosub middle                      ; Goto the subroutine middle
7     elseif b1>20 then                 ; If the value of b1 is >20 then
8           gosub big                   ; Goto the subroutine big
9     endif                             ; Stop testing b1
10    goto main                         ; Return to the start
11
12    small:                            ; The subroutine small
13    high B.4
14    low B.5
15    low B.6
16    return                            ; Goto line 5
17    middle:                           ; The subroutine middle
18    high B.5
19    low B.4
20    low B.6
21    return                            ; Goto line 7
22    big:                              ; The subroutine big
23    high B.6
24    Low B.4
25    Low B.5
26    return                            ; Goto line 9
```

BASIC *elseif* command.

Now try this

Explain the use of the *switch* flow control functions on a platform of your choice.

Control of program sequence: switch 2

This page revises the *switch case* function in BASIC in the Arduino™ platform. Arduino™ *switch* and BASIC *elseif* are revised on page 216.

Using C to perform *switch case* control in BASIC

The program shows how the *switch case* function is used to control the sequence the program follows.

```
void loop() {
  int sensorReading = analogRead(A0);
  //The map function below converts all the possible values from the ADC to one of 3 values 0, 1 or 2
  int range = map(sensorReading, sensorMin, sensorMax, 0, 2);
① switch (range) {
②  case 0:                                    // do this if the sensor reading is small
     digitalWrite (smallValue, HIGH);         // This turns on the small LED
     digitalWrite (middleValue, LOW);         // This turns off the middle LED
     digitalWrite (bigValue, LOW);            // This turns off the big LED
     break;                                   // This returns the program to the analogueRead function
③  case 1:                                    // do this if the sensor reading is in the middle
     digitalWrite (smallValue, LOW);          // This turns off the small LED
     digitalWrite (middleValue, HIGH);        // This turns on the middle LED
     digitalWrite (bigValue, LOW);            // This turns off the big LED
     break;                                   // This returns the program to the analogueRead function
④  case 2:                                    // do this if the sensor reading is big
     digitalWrite (smallValue, LOW);          // This turns off the small LED
     digitalWrite (middleValue, LOW);         // This turns off the middle LED
     digitalWrite (bigValue, HIGH);           // This turns on the big LED
     break;                                   // This returns the program to the analogueRead function
  }
  delay(1000);                                // delay in between reads for stability
}
```

> The program above shows how the *switch case* function is used to control the sequence the program follows. ① is where the switch **function** is called. ②, ③ and ④ show the **three cases** that could be followed.

Help files

System designers need to be able to extract information about using this type of command from the **help files** available for the microcontroller platform. Part of the help file for the **switch** command is shown below.

Like *if* statements, *switch...case* controls the flow of programs by allowing programmers to specify different code that should be executed in various conditions. In particular, a *switch* statement compares the value of a variable to the values specified in *case* statements. When a *case* statement is found whose value matches that of the variable, the code in that *case* statement is run.

Now try this

Determine why the break; command is needed for the program. Information about this can be found within the reference section of the learning part of the Arduino™ website.

Structured program design

You can use pseudocode, flowcharts and decision tables to structure a microcontroller program.

Pseudocode

Pseudocode allows you to plan the structure of a microcontroller program with a focus on the **operating requirements** rather than the programming language. It is written in simple **everyday language** such as:
Wait until a switch is pushed

The **equivalent function** written in C is:
If digitalRead(7, 1) then

There is no formal definition for the structure of pseudocode but common notations include:

Input, output, while, for, repeat-until and if-then-else

Flowcharts

There is no official structure for flowcharts. Flowcharts use **symbols** specified by BS4058 or ISO5807. The shape of each symbol is related to a **specific command** or **function**.

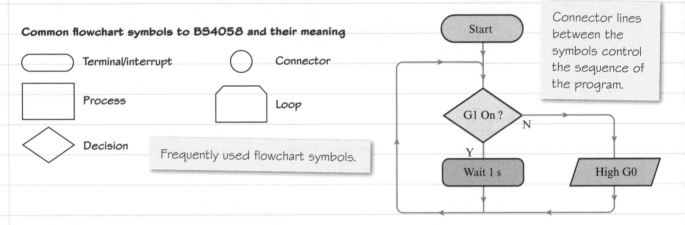

Common flowchart symbols to BS4058 and their meaning

Terminal/interrupt   Connector

Process   Loop

Decision

Frequently used flowchart symbols.

Connector lines between the symbols control the sequence of the program.

Start → G1 On ? → Y → Wait 1 s   N → High G0

Decision tables

A decision table **models the rules** that a microcontroller program must follow. A decision table consists of **four sections**.

1️⃣ **Conditions** are the inputs to the system.

2️⃣ **Actions** are the outputs from the system.

3️⃣ **Condition alternatives** show all the combinations of inputs that are possible.

4️⃣ **Action entries** show what outputs are activated for each combination of inputs.

1️⃣ Conditions	3️⃣ Condition alternatives
2️⃣ Actions	4️⃣ Action entries

Decision table for an alarm system

			Rules		
1️⃣ Conditions	3️⃣	Alarm turned on	Y	Y	Y
	3️⃣	Sensor activated		Y	Y
	3️⃣	Day time		Y	
	3️⃣	Night time			Y
2️⃣ Actions	4️⃣	Sound alarm		X	
	4️⃣	Flash light		X	X

The table shows that if the alarm is turned on, and a sensor is activated, and it is daytime, an alarm will sound and a light will flash. If the same events happen at night only the light will flash.

Now try this

Construct a flowchart to run an alarm system similar to the one described above. Refer to flowchart symbols approved by BS4058.

Number systems: decimal to binary

This page revises how to convert between the decimal and binary systems, and revises the terms **bits** and **bytes**.

Bits and bytes

- Microcontrollers operate using the **binary**, or base 2, number system.
- A **binary** number is represented by a series of **binary digits**, which is where the term **bits** originates.
- A **bit** can store only **two** possible values, 0 or 1.
- When **eight bits** are written in sequence they become known as a **byte**.
- A **byte** can store **256 different** unique values.
- The table below shows the **equivalent decimal value** of each bit in a byte.

	Most significant bit							Least significant bit
	bit 7	bit 6	bit 5	bit 4	bit 3	bit 2	bit 1	bit 0
Binary	2^7	2^6	2^5	2^4	2^3	2^2	2^1	2^0
Decimal	128	64	32	16	8	4	2	1

The most **significant** bit is **bit 7**, which has the value of 2^7 or $2 \times 2 \times 2 \times 2 \times 2 \times 2 \times 2$ or 128.

Consider the number 156. In the **decimal** system 156 means $(1 \times 10^2) + (5 \times 10^1) + (6 * 10^0)$.

In the **binary** system the number 156 would be the same as $(1 \times 2^7) + (1 \times 2^4) + (1 \times 2^3) + (1 \times 2^2)$ or 10011100.

	bit 7	bit 6	bit 5	bit 4	bit 3	bit 2	bit 1	bit 0
Value	2^7	2^6	2^5	2^4	2^3	2^2	2^1	2^0
	1	0	0	1	1	1	0	0

	Decimal number	Remainder
	156	
Divided by 2	78	0 (last digit of binary number)
Divided by 2	39	0
Divided by 2	19	1
Divided by 2	9	1
Divided by 2	4	1
Divided by 2	2	0
Divided by 2	1	0
Divided by 2	0	1 (first digit of binary number)
Binary number	10011100	

Converting a decimal number to a binary.

Decimal to binary conversion

1 Write down the number to be converted (e.g. 156).

2 Then divide by 2, and write the answer below (78).

3 If there is no remainder, write 0 on the right.

4 If there is a remainder, write 1 on the right. Repeat the process until you arrive at zero.

5 Finally, write out the remainders from bottom to top to give the binary conversion of the decimal number.

Now try this

Convert these decimal numbers to binary:
131, 172, 94.

Unit 6
Content

Number systems: binary to decimal

This page revises how to convert between the **binary** and **decimal** systems and revises the difference between **parallel** and **serial** connections.

Binary to decimal conversion

1 Draw the table that shows the powers of 2 and their decimal equivalent.

2 Next, enter the binary number below this table.

3 Wherever a binary 1 has been written, copy the decimal value above the bit into the cell below.

4 Finally, add up the decimal numbers. This gives the decimal equivalent of the binary number.

First ➡

2^7	2^6	2^5	2^4	2^3	2^2	2^1	2^0
128	64	32	16	8	4	2	1

Second ➡

2^7	2^6	2^5	2^4	2^3	2^2	2^1	2^0
128	64	32	16	8	4	2	1
1	0	0	1	1	1	0	0

Third ➡

2^7	2^6	2^5	2^4	2^3	2^2	2^1	2^0
128	64	32	16	8	4	2	1
1	0	0	1	1	1	0	0
128			16	8	4		

Last ➡ 128 + 16 − 8 + 4 = 156

> Converting binary numbers to decimal.

Parallel and serial connections

Serial and **parallel** transmissions are two methods of sending binary data along wires, or connections, between components.

- A **serial** connection sends the data along a single wire.
- A **parallel** connection sends the data along multiple wires.

The diagram shows a **parallel** connection with eight wires, which allows **one byte**, or eight bits, to be sent at the same time. Parallel transmission can send data **faster** than serial, but parallel connections are more complex and costly to manufacture.

> Parallel and serial transmission

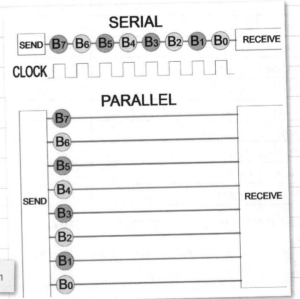

Now try this

Convert these binary numbers to decimal: 10001101, 00111110.

Project analysis

To help make your microcontroller system successful, you must make an accurate **analysis** of the **client brief**.

Example of a design brief

Scenario: A garage owner has asked you to develop a prototype alarm system to make the garage more secure. This will help to decide if a complete working system is viable.

Client brief: TL Garage is a car repair garage. The garage uses a range of expensive equipment that the owners are worried could be stolen. Because the garage is in a residential area, the local council will not allow a burglar alarm to be installed that would disturb any neighbours when it is dark. The owner of the garage has identified three main areas that they would like to be monitored by an alarm system. The entrance to the garage is between two walls, 3 m apart. The garage owner would like a warning to be displayed if a car or person comes into this entrance. There is a rear door to the garage. The garage owner would like an audio and visual warning to be triggered if the door opens when the alarm is turned on. There is a secure storeroom. The garage owner wants the entry door to be alarmed at all times unless she specifically turns the alarm off. The garage owner has allowed you 12 hours to complete your investigation and has allocated a budget of £20 for you to purchase any extra equipment that you may need.

Analysis

As system designer, first make sure that you understand what you are required to achieve.

This involves extracting the **critical information** from the **client brief**. You must take into account what the system is required to do and any restrictions that are imposed.

- One approach is to identify the **technical elements** of the design brief and classify them into **requirements** and **restrictions**. In the example above, requirements are highlighted in green, and restrictions are in red. Other key pieces of information are underlined.

- As well as information that is explicitly provided by the design brief, you may need to consider if any relevant information is **missing** from the brief and what **assumptions** you need to make.

Production of a technical specification

From the analysis of the design brief you then need to produce a **technical specification**. The specification will need to consider two main components of the project:

1 What does the **software** need to do?

2 What does the **hardware** need to do?

A critical element of the technical specification will be to allocate **measureable performance criteria** to all of the requirements given in the design brief.

Creation of a test plan

From the technical specification a **test plan** can be developed. The plan should clearly set out:

- **What** is going to be tested?
- **How** is it going to be tested?
- **When** is it going to be tested?
- **What** are the pass / fail criteria?
- **How** will the results of the test be used?

At this stage you may realise that you have missed something from your analysis or specification. This would then prompt the **analysis > specification > test plan** cycle to be repeated until you are satisfied that all aspects have been considered.

Now try this

Produce a design brief that includes requirements and restrictions for a situation of your choice.

System design and program planning

System design and program planning requires a structured approach.

Selection of control hardware

1 Use your **time efficiently**, as the completion of most projects is time critical.

2 Select **hardware platforms** that you have experience using, provided they are able to meet the needs of the client brief.

3 Select and justify the **input / output** devices to be used, using ones that you are already familiar with. Although the specification of the components needed will vary from project to project, the method of interfacing them to the microcontroller and the software functions needed to operate them should remain reasonably constant.

4 Consult suppliers' catalogues to find **available products**.

5 **Refine your choice** by confirming more specific factors, such as:

- Will the device **perform** as required? The product data sheet should be able to answer this.
- Will the device **interface** with the hardware platform? Again, data sheets will be useful.
- Can the cost of the device be met by the **budget**? This will probably need to be revisited as you develop your design.
- Will the device be **available** to use within the time scales of the project?

Often, sourcing components from other countries can reduce their cost, but they may take too long to arrive. All of the factors above need to be balanced to arrive at the most appropriate choice of device to use.

Design of input/output circuitry

1 Consider the **physical connections** available on the microcontroller itself.

2 Use manufacturers' data sheets to find details of what each **connection pin does**.

3 Determine what **connections** will be required between each component to allow the system to function as required.

You can find PIC® pin out sheets online.

Simulation

Some CAD systems, such as Circuit Wizard, allow you to **simulate** the interaction between the user-written software and the hardware system you have designed.

Design of the program structure

1 Choose the most appropriate **structure** to solve the particular situation the design brief relates to.

2 If the design brief consists of **multiple** different elements it can help to consider them as **separate tasks**, with their own demands and solutions.

3 Once you have **implemented** and **tested** these individual solutions, you can **assemble** them into the complete system.

4 At this stage, **assess the interactions** between the various components and **amend** if necessary.

Links See page 218 covering a range of techniques for designing program structures.

Now try this

Identify any input/output components listed in this Revision Guide that you need to be familiar with. Make sure you have a working knowledge of their attributes before the controlled assessment task.

System assembly, coding and testing

When, as the system designer, you are ready to start constructing the hardware elements of the project, you will need to access information about how to connect the components.

Accessing information

- The **breadboard view** is easy to understand when few components are involved.
- The **schematic diagram** becomes more useful as the complexity of the circuit increases.
- **Text instructions** are useful on their own or in support of diagrams.

> Information may be available as an image, a schematic or text.

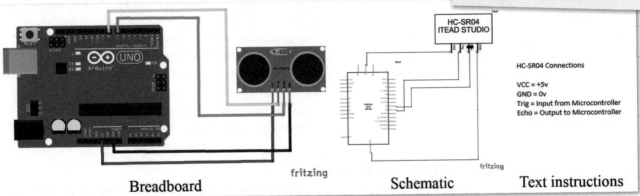

HC-SR04
ITEAD STUDIO

HC-SR04 Connections

VCC = +5v
GND = 0v
Trig = Input from Microcontroller
Echo = Output to Microcontroller

fritzing

Breadboard   fritzing   Schematic   Text instructions

Assembly and connection of physical components

1 **Assemble** each of the inputs, and associated outputs, as separate circuits on breadboard.

2 Ensure the **connections** are correct, then test their performance using the microcontroller. When the circuit works, **record** the connections accurately.

3 When all the **separate** inputs and outputs work individually, **test** the **complete system** to ensure there is no **unexpected interaction** between components and/or software.

4 You can then create the **permanent circuit**.

Program authoring

You can convert your intentions into the **working program**.

- Make **adjustments** until the program performs as expected.
- Add appropriate **annotation** as you construct the code.
- Check the **formatting** of the program to ensure that a third party could update the code at a later stage if required.

Developmental testing and refining

When **testing** and **refining** the microcontroller system you should consider the benefits of changing just **one parameter** at a time. This will allow the effects of the change to be more clearly understood.

> The annotations should explain to a reader the 'how and why' of the code.

Now try this

The breadboard diagram above uses different coloured wires to connect components to the microcontroller. Explain why this may be an advantage for the system designer.

System testing and operation

When you have completed the development of the individual elements that make up the system, you must combine and test them to form the complete working system.

Operation of the finished system

Check that you have not introduced any unintentional changes while combining the individual elements that make up the system.

- Consider combining just **two elements** at a time (rather than combining all elements in a single stage) and testing the results to ensure the system still works as expected.
- The **advantage** of this approach is that you can more easily identify any interactions between elements that can cause problems.
- A **disadvantage** is that it may increase the time required to complete the working system.

Carry out the test plan and record results

Follow these general principles to test your microcontroller system:

- Plan to test **all** the features of the system in the time available.
- Subject the most **complex** elements of the program to the most vigorous testing.
- Test analogue sensors across the **full range** of the variable they will measure. This should include at least the **maximum** and **minimum** values, with an appropriate number of intermediate readings.
- Where possible, record **quantitative measurements** of the variables as well.
- Record the **purpose** of each test, the **data** used to perform the test and the **expected** and **actual results**.

Description of the system operation

Having completed the testing of the project, describe how the system operates. This explanation should provide the client with specific details of the **components** used, what **function** they perform and how the microcontroller **interacts** with them. An example is shown below for an automatic night light.

Sample response extract

The level of light is measured using a Silonex light-dependent resistor, model NORPS12. This LDR has a resistance of over $1\,m\Omega$ in the dark and around $1\,k\Omega$ in bright light. Using a multimeter, it was determined that the resistance of the LDR was around $100\,k\Omega$ when the light level was 40 lux. Through experimentation this was determined to be when the night light should turn on.

The LDR is connected to the analogue to digital convertor on pin 23(A0) of the Arduino™ Uno. At this light level the ADC gives a value of 180. The reading of the ADC was stored in the integer variable called LightLevel. The program performs a conditional test on this variable to determine if the value is above or below 180. The program reads the value of the LDR every 10 seconds. When the value is below 180; i.e. dark, the digital output pin 15(PB1) is turned on. Connected to this pin is a light-emitting diode, specifically a TruOpto OSW5DKA132A-ST 10mm White Diffused LED. The LDR and LED are arranged in such positions that light from the LED does not reach the LDR. This prevents the continuous oscillation of the state of the LED. Details of the connections between the components and the microcontroller are shown by the schematic circuit diagram.

Analysis of test data

Analysis of test data should:

- **Confirm** the system operates as expected or
- **Identify** the reason(s) for any **divergence** from the required system performance.

Evaluating the completed system

As system designer, your evaluation of the project should:

- Identify both **positive** and **negative** attributes of the system.
- Provide the client with an **objective overview** of how well the system performs.
- Suggest areas for **future development**.

Now try this

Consider one feature of a system you have developed and describe its operation.

Production of evidence

Below is a guide to the evidence required for the assessment of the unit.

Production of a technical specification
- Interpret the design brief to produce a set of operational requirements.
- Consider how users of the system will interact with it. This interaction should enhance the user experience.

▼

Production of a test plan
- Consider from the outset how to confirm that the system meets the demands of the client brief.
- Devise a logical sequence of tests with objective success criteria predetermined to compare system performance to. Also take into account how the system will perform when unexpected events occur.

▼

Selection and justification of input and output devices
- Select input and output devices to meet the demands of the system.
- Provide accurate descriptions of the input and output devices that justify the reasons for their selection.

▼

Production of system connection diagrams
- Develop diagrams and associated annotation to detail how input and output components will interface with the microcontroller.
- Use technical descriptions and drawings that comply with appropriate standards to convey the information.

▼

Production of the initial design for program structure
- Break the program into components that link together logically to form the complete system.
- The initial program design should take into account how to handle unexpected events.

▼

Production of annotated copies of all code
- Use a range of constructs to produce efficient code with the ability to handle unexpected events.
- Annotate the code in a consistent manner that demonstrates your understanding.
- Structure the program so that a third party could update it effectively.

▼

Recording of test data and analysis
- Record the results of a range of tests carried out on the system. The tests should include the simulation of unexpected events.
- Compare the results of the tests to the demands of the client brief and make judgements about the extent to which the program meets the client requirements.

▼

Production of an audio-visual recording of system operation
- Record, using audio-visual equipment, the system functioning. The recording should demonstrate how the system provides an enhanced user experience and handles unexpected events.
- The recording should also show a thorough understanding of the relationship between the hardware and software, using accurate technical terminology.

▼

Production of a structured project log
- Check that your records show that you followed a structured approach to the development process.
- Use accurate technical terminology to describe the activities carried out.
- Check that entries in the project log are concise and justify any changes made to original design intentions.
- At the end of each assessment period, plan and prioritise action points for the next session.

- -

Now try this

Produce a simplified version of the guide (for quick reference) that includes only the topics not the descriptions.

Your Unit 6 set task

Unit 6 will be assessed through a task, which will be set by Pearson. In this assessed task you will need to develop a prototype microcontroller system to solve an engineering problem.

Revising your skills

Your assessed task could cover any of the essential content in the unit. You can revise the unit content in this Revision Guide. This skills section is designed to **revise skills** that might be needed in your assessed task. The section uses selected content and outcomes to provide an example of ways of applying your skills.

Reading a task brief
See page 227

Creating and monitoring a task plan, recording and justifying changes
See pages 228–231

Analysing a brief for product requirements and completing a test plan See pages 232–233

Set task skills

Selecting input and output devices, formulating a system design, proposing connections and planning a program structure
See pages 234–236

Recording a system in operation and test outcomes, with commentary
See pages 240–241

System testing and results analysis
See pages 238–239

System assembly and programming
See page 237

Workflow

The process of developing a prototype microcontroller system to solve an engineering problem might follow these steps:

✓ Read a task brief.

✓ Create a project time plan and use it to monitor progress and record changes.

✓ Analyse the brief for product requirements and use it to generate a technical specification for the system, completing a test plan for suitability of the final solution.

✓ Select and justify input and output devices and formulate an initial design system, proposing system connections and planning a program structure.

✓ Assemble your hardware, author your program and annotate your code.

✓ Test your system against your plan and record the outcome of each test. Analyse test results and evaluate your system for conformance against the technical specification.

✓ Make an audio-visual recording to demonstrate your solution in operation and test outcomes, with explanatory commentary.

Check the Pearson website

This section is designed to demonstrate the skills that might be needed in your assessed task. The details of the actual assessed task may change so always make sure you are up to date. Check the Pearson website for the most up-to-date **Sample Assessment Material** to get an idea of the structure of your assessed task and what this requires of you.

Now try this

Visit the Pearson website and find the page containing the course materials for BTEC National Engineering. Look at the latest Unit 6 Sample Assessment Material for an indication of:

- The structure of your set task, and whether it is divided into parts
- How much time you are allowed for the task, or different parts of the task
- What briefing or stimulus material might be provided to you
- Any notes you might have to make and whether you are allowed to take selected notes into your supervised assessment
- What you might need to take into the assessment, e.g. a calculator
- The activities you are required to complete and how to format your responses, e.g. on a computer with appropriate hardware and software; the specification and length required for the audio-visual recording.

Reading a brief

Here are some examples of skills involved when reading a scenario and client brief.

Scenario and client brief

The scenario and client brief are used as examples to show the skills you need. The content of a task will be different each year and the format may be different.

When reading a scenario and client brief:

- read it several times
- underline key information to help you focus on the most important aspects.

For the below scenario and client brief, the task might be, for example, to design, assemble, program and test a quiz game to meet the requirements of the client brief, following appropriate development processes and using a microcontroller.

Scenario

You are employed as a software engineer by a games manufacturer. The games manufacturer wants to develop a quiz game that follows the format of the TV game 'On the Buzzer'. You have been presented with the client brief to develop a system that will tell the quiz master which person was first to push the buzzer.

> Read the scenario and client brief carefully. As system designer, you need to understand what you need to achieve.

Client brief

Tooley-Lober Games are a public limited company <u>based in the UK</u>. They sell a range of traditional board games and <u>children's toys</u>. The market for these products has become saturated and they are looking for ways to develop new income streams.

The company has agreed to a licensing arrangement with the TV programme 'On the Buzzer' to produce a game that allows <u>up to four players</u> to <u>answer questions</u> of the same style as the program.

The marketing director of the company has determined that an electronic method of indicating which player was first to push their button is likely to increase sales as it will become a unique selling point.

When the <u>quiz master starts reading the question</u> the <u>first player to press their button</u> will have the opportunity to provide an answer. Once the first person has pressed their button the <u>other player's buttons must become inactive</u>.

Pressing the button <u>must clearly indicate to the quiz master which player was the first to press the button</u>.

Having pressed the button <u>the player is allowed 10 seconds</u> to answer the question. The system must indicate when <u>this period of time has elapsed</u>.

> Identify the operational and technical elements of the brief. What does the software need to do? What does the hardware need to do?

The quiz master must be able to <u>control when the players' buttons are active</u> and be able to <u>reset the system</u> when an answer has been provided, or the <u>time available to answer</u> has run out.

Client name: Tooley-Lober Games

- Task planning and system design changes made during the development process
- Interpret a brief into operational requirements
- Design a test plan based on operational requirements
- Select and describe appropriate input/output components and how they will work together
- Design the program structure

- Produce a functional system
- Annotate a program or code to demonstrate understanding
- Test the system and analyse the outcomes from testing
- Produce an audio-visual recording of the system in operation of no longer than three minutes

🔗 Links   To revise analysis of the brief, look at pages 221 and 232.

Now try this

Suggest two or three factors you will need to consider when designing a product that might be used by children.

Creating a task plan

Here are some examples of skills involved when creating a short project time plan. Use the plan to **monitor** your progress and **record** changes to your original design, providing details of any **issues** you encounter and **justifying the solutions** you discover.

PROJECT TIME PLAN FOR THE TASK

Task planning and system design changes, e.g.:
- structured approach to carrying out an iterative development process, in appropriate order, using technical terminology
- justification of changes with logical and prioritised action points
- I will review previous activity and plans, making notes and, if appropriate, take photographs, of significant events, recording 'how' and 'why'. I will record progress and produce well-defined, logical and prioritised action points to plan the next stage.

> Consider what each stage requires and how it breaks down into specific tasks, allocating time to each part. You should demonstrate a **logical** and **iterative** approach to the design process. Make sure that your action points show forward planning clearly linked to the specific context, with consideration of what has happened in the previous session.

Analysis of the brief, e.g.:
- analyse the scenario, background information and product requirements
- use the analysis to generate a technical specification and interpretation of the brief into operational requirements
- test plan with parameters that are designed to confirm a fully functioning system including unexpected events
- I will...

> Make sure you interpret the brief into a set of cohesive operational requirements, not just a repeat of the client brief, considering also enhanced user experience.

System design, e.g.:
- hardware input and output device selection, appropriate for the operational requirements
- description of microcontroller interface to hardware
- use of technical terminology and industry standard conventions
- program design linked logically and coherently, including the handling of some unexpected events
- I will...

> Ensure your ideas are feasible and fit for purpose, considering unexpected events.

> Use appropriate technical terms for your annotation.

System assembly and programming, e.g.:
- program quality – a range of appropriate constructs that have been used correctly...
- annotation to be...
- the program to be well-organised, structured and formatted so that...
- I will...

> Make sure you optimise your solution, ensure the proposal is informed and justify your choices.

System testing and results analysis, e.g.:
-
-
- I will...

Complete the detail for the plan above, then consider the time to allow for each of the stages. Use the information on timing and marks from the latest Unit 6 Sample Assessment Material and Mark Scheme on the National Engineering page of the Pearson website, to guide you.

System in operation, e.g.:
-
-
- I will...

> You should give a balanced and thorough commentary, with a sound rationale.

Monitoring progress

Here are some examples of skills involved when you have completed an initial plan and **monitor** your progress towards the **deadlines** you have set.

Project log

Use a project log for each stage of design to record your:

- general comments
- issues encountered and solutions, with justification
- action list for the next stage.

Sample response extract

General comments

13/6/17

Factors that need to be considered:

The project revolves around a game, which could mean children will use it. What does this mean?

The product should be easy to use, it should appeal to children/adults.

20/6/17

The project will use a PICAXE® 28X2 chip programmed using BASIC. This is available, I am familiar with its use and it provides sufficient input / output connections.

In this extract recording progress, two sections of the task brief have been **recorded**. By stating the date, you can follow an appropriate **order**.

The record reflects the **context** in which the microcontroller system operates, as provided in the **scenario**.

The record reflects what the microcontroller system is required to perform, as outlined in the **client brief**. Your technical specification should detail how the customer's operational requirements will link together to create a functioning and coherent **solution**.

Sample response extract

Issues encountered and solutions, with justification

I had been waiting for a yellow switch part number 34-123A to arrive but was informed by an email from the supplier that it would not be available for another 10 days. I do not think I can wait this long so I will have to think about what to do.

In this extract, the issue has been identified but there is **no** solution, and the record is **not** concise.

Improved response extract

Issues encountered and solutions, with justification

Due to unavailability of the switch I planned to use, I will replace it with one that I already have access to. This will allow me to continue testing and development.

In this extract from an improved response, the issue encountered is **stated** and the solution is **justified**.

Make sure that you complete log entries for each activity. This may mean you need to move on from some tasks even when you want to do more work on them. Make sure your next points of action are well-defined, logical and prioritised.

Now try this

Compile a list of the resources available for you to use at your centre. It may be useful to include a focus on the availability of input and output devices, as these may be critical components. Having this information in advance will help you to be prepared and use your time well.

In the example above, a delay was caused by the unavailability of a component. This type of delay can be eliminated by making use of resources that are already available.

Monitoring changes 1

This page and the next contain examples of learner evidence with suggestions about how they could be improved. Four versions of a program are given below.

Sample response extract

In this extract, **annotations** allow a **third party** to **interpret** the code.

```
39          Quiz_Master:                                      'Wait for the question to be answered
40  ①      if pinC.5=0 then Quiz_Master                      'Turn off the players light
41          low C.0,C.1, C.2, C.3                             'If the answer is correct show the correct light
42          if pinC.6=1 then high B.6 return else endif       'If the answer is incorrect show the incorrect light
43          if pinC.7=1 then high B.7 return else endif

39          Quiz_Master:                                      'Wait for the question to be answered
40  ②      if pinC.5=0 then Quiz_Master                           'Turn off the players light
41          low C.0,C.1, C.2, C.3, C.5                        'If the answer is correct show the correct light
42          if pinC.6=1 then high B.6 return else endif       'If the answer is incorrect show the incorrect light
43          if pinC.7=1 then high B.7 return else endif

41          Quiz_Master:                                      'Wait for the question to be answered
42  ③      if pinC.5=0 then Quiz_Master                      'Turn off the players light and quiz masters buttons
43          low C.0,C.1,C.2,C.3                               'If the answer is correct show the correct light
44          if pinC.6=1 then high B.6                         'Display the correct answer light for 5 seconds
45          pause 5000                                        'Go back to reading players buttons
46          return else endif                                 'If the answer is incorrect show the incorrect light
47          if pinC.6=0 then high B.7                         'Display the incorrect answer light for 5 seconds
48          pause 5000                                        'Go back to reading players buttons
49          return else endif

42          Quiz_Master:                                      'Wait for the question to be answered
43  ④      if pinC.5=0 then Quiz_Master                      'Turn off the players light and quiz masters buttons
44          low C.0,C.1,C.2,C.3                               'If the answer is correct show the correct light
45          if pinC.6=1 then high B.6                         'Go back to reading players buttons
46          return else endif                                 'If the answer is incorrect show the incorrect light
47          if pinC.6=0 then high B.7                         'Go back to reading players buttons
48          return else endif
```

There is **evidence** of **iterative** development from version 1 to 4.

When logging changes, you need to make sure that you **justify** the changes made to your program to demonstrate your ability to apply logical reasoning.

Microcontroller hardware and software

The only acceptable microcontroller hardware and programming languages for Unit 6 are shown on page 172. Always check you have the up-to-date details by looking at the specification on the Pearson website. In this skills section, examples of developing a solution in response to the brief are given using selected hardware and software. If you are developing your solution using a different microcontroller platform, you should follow the same approach. Take the software instructions and apply them so that you can perform the same functions in the microcontroller platform that you are using.

Now try this

Look at the changes that have been made to the examples above. Write a justification for these changes.

Monitoring changes 2

Here are some examples of suggested changes for improvement that could be logged in relation to completion of a test plan to test the suitability of the final solution. You will also need to analyse results to assess how far the results show that the system meets the client brief.

Sample response extract

Here are some extracts from test plan templates. A technical specification should be brought together from a client brief and be used to help complete a test plan to test the suitability of the final solution. Changes should be logged and justified.

These tests were planned before the system was implemented.

Test no.	Purpose of test	Test data	Expected result	Actual result	Comments and justification
1	To determine if the system performs as expected when a single player button is pushed.	Player 1 button is pushed.	A buzzer will sound if a player uses a button. A light will come on, showing which player pushed. The QM will determine if the answer is correct and play will continue.	Just touching the wire to the button worked the same as pushing the switch. Play continued with intervention from the QM.	I had forgotten some inputs could be touch sensitive, so the hardware interface connections need amending.

These tests were devised in response to the system being modified to solve problems

| 1a | To determine if the system performs as expected when a single player button is pushed. | Player 1 button is pushed. | The program should function only when a player actually pushed their button. | Problems with touch sensitive inputs were no longer an issue. | This allows me to develop the rest of the system. |

This extract from Test 1a indicates that the changes made solved the problem but there is **no reference** to how this was achieved.

Improved response extract

These tests were devised in response to the system being modified to solve problems

Test no.	Purpose of test	Test data	Expected result	Actual result	Comments and justification
1a	To determine if the system performs as expected when a single player button is pushed, using the modified connections shown in schematic diagram 2.	Player 1 button is pushed.	The program should function only when a player actually pushed their button and not as a result of wires just being touch sensitive.	Problems with touch-sensitive inputs were no longer an issue.	This allows me to develop the rest of the system, as the results of this test now match what is required.

This improved extract presents the same results but **supports** and **justifies** the comments with reference to other related **evidence**.

Now try this

This photograph shows a visual record of the system in operation, as part of the evidence needed for test outcomes. It includes a panel that holds the lights that come on when a player pushes their button and a light to indicate if the answer is correct or not. The input/output devices have been labelled to show which feature of the program they relate to. Suggest how the labels could be improved to make it clearer for the viewer what is being demonstrated.

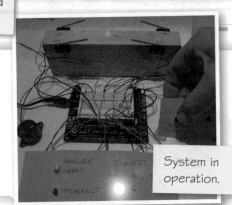

System in operation.

Analysing a brief for product requirements

You need to consider the operational requirements arising from a client brief.

Sample notes extract

Client brief

- UK based
- up to four players
- QM starts reading the question
- first player to press their button
- other players' buttons must become inactive
- must clearly indicate to the QM which player was the first to press the button
- the player is allowed 10 seconds
- system indicates when this period of time has elapsed
- QM controls when the players' buttons are active
- QM resets the system

> These operational requirements have been brought together from the client brief on page 227. Underlining key parts of the client brief can help you focus on the most important aspects when determining how you will design the product.

Sample response extract

> This extract from a sample response shows how a client brief is starting to be **interpreted** into a set of **operational requirements**. It is a sensible approach to start with general points such as 'something will need to happen' and to **refine** them as time progresses.

Factors that need to be considered	What does this mean?
The project revolves around a game, which could mean children will use it.	The product should be easy to use and should appeal to children and adults, if possible. Ergonomically, a range of hand sizes should be able to work the controls.
The company is based in the UK so the product should comply with UK standards.	BS EN 71 applies to electronic toys so the content of this should be observed.
No detail is provided about the desired size of the product.	Most board games are a similar size so it would make sense to ensure the product fits in a standard box.
There are up to four players and quiz master. Push button inputs are required.	Probably players will sit around a playing board so each one will need a separate button. The quiz master will need information from the players' buttons.
When the quiz master starts reading, players' buttons must work.	This suggests they must not work before then.
Once one button has been pushed other players' buttons must stop working.	If the buttons are pushed at very similar times the system will need to work quickly.
A 10 second timer is needed.	At the end of 10s something will need to happen, maybe a light or sound?
The quiz master must be able to control the system.	All the players' buttons must connect to the quiz master's device.

Now try this

Suggest a possible output device that could be used to signal the end of 10 seconds.

> Remember that there is a time limit for developing the complete system. You will need to balance the features available on the output device with the time it will take to incorporate it into the system. The KISS principle is something to consider (Keep It Simple, Stupid).

Completing a test plan

From the analysis of a brief, you need to produce a **technical specification** and complete a plan for **testing** the suitability of the final solution.

Sample response extract

🔗 **Links** To revise system testing, look at pages 238–239.

Technical specification

- The project will use a PICAXE® 28X2 chip programmed using BASIC. This is available, I am familiar with its use and it provides sufficient input / output connections.
- The system will be battery powered. This will reduce cost, increase portability and ensure safety.
- Push to make switches will be used by the players to show they want to answer a question. This will stop all other players' buttons. The number of players can vary so the system should not depend on all four players pushing switches.
- Lights will be used to indicate which player has pushed the button first.
- Sounds will be used to indicate when 10s has elapsed. The time will start as soon as a player pushes a switch.
- The quiz master will use switches to control the system.
- The system should be designed to fit into a conventional size box for a board game.

Remember that you can **amend** or **add** requirements as the project develops.

Sample response extract

Test no.	Purpose of test	Test data	Expected result	Actual result	Comments and justification
1	To determine if the system performs as expected when a single player button is pushed.	One player's button is pushed.	A buzzer will sound if any player has pushed their button. A light will come on to show which player pushed. The QM will determine if the answer is correct and play will continue.		
2	To determine what happens if two or more players push their buttons at the same time.	I will connect a switch so all four inputs are triggered at once.	There should be a random outcome for which player is allowed to answer the question.		
3	To determine how long the system takes to react to a button being pushed.	Time between button pushed and light on	Less than 1 second		
4	To determine the accuracy of the time delays built into the system.	Duration of LEDs and buzzer staying on	Within 10% of expected time		
5	To determine if the correct sequence of lights and sounds occurs when the quiz master determines if the answer is correct or not.	All combinations of players and right / wrong answers	Accurate indications given all the time.		

Here is an extract from a test plan template, with an example of planning how to test the system, in appropriate detail. You can also consider responses to unexpected events.

Compare features on related products. Might they generate suitable ideas to enhance this product?

Now try this

Suggest an addition to the system that considers enhanced user experience.

Formulating a system design

When you have analysed a brief, you need to select your **input** and **output** devices and formulate an initial design for the system. Below are examples of the components selected and some reasons for selecting the component.

Sample response extract

I will use the 28X2 microcontroller.

Improved response extract

I will use the 28X2 microcontroller. I have used this before and it is available. It has the capacity to connect more input / output devices than I plan to use. Having the ability to connect more devices will allow me to add enhanced features if I have sufficient time.

 Links To revise system design and program planning, look at page 222.

In this extract from a response, appropriate hardware has been selected but the choice has **not** been justified.

In this improved extract, comments are given that **justify** the selection of components. Reference is made to the **possibility** of adding enhanced user experience. You need to focus on the **essential requirements** before adding additional features.

Sample response extract

I will use 37.5 mm lever micro-switch player inputs. These switches are hardy enough to withstand some abuse by the kids playing the game.

In this extract from a response, the appropriate hardware has been selected but appropriate technical terminology has **not** been used.

Improved response extract

I will use micro-switches for the player input. This is because they are momentary action switches that do not latch when pushed by the player. They have a mechanical life expectancy of 10 million operations, so durability would not be an issue.

This improved extract makes use of **technical terminology** and is an appropriate **style** for a technical document. It **justifies** the selection of hardware and uses industry standard **conventions**.

This extract shows a list of components required to construct the proposed system, using a typical industry standard convention. Using part numbers means the specific components are clear.

Sample response extract

Component description	Part number	No.	Cost	Total
Micro-switch 37.5 mm lever	2470-156	4	£0.63	£2.52
Miniature toggle switch SPST On–Off	0125-156	1	£0.69	£0.69
RED miniature red push switch	0030-156	1	£0.48	£0.48
5 mm ultra red LED diffused 320 mcd	0486-157	1	£0.18	£0.18
5 mm orange LED diffused 320 mcd	0858-157	4	£0.08	£0.32
5 mm super bright green LED diffused 320 mcd	0884-193	1	£0.08	£0.08
6 V electronic buzzer with 20 cm lead	3588-601	2	£1.05	£2.10
PICAXE® 25×2 chip	5002-698	1	£6.32	£6.32
PICAXE®/Genie® compatible 28-Pin PIC Prototype PCB kit	7127-658	1	£3.59	£3.59
			Total	£16.28

Now try this

The proposed system makes use of LEDS. Write a short paragraph to justify their selection in the style shown above.

Proposing system connections

Here is an example of the **initial design proposals** for the **hardware** components of the project.

Sample response extract

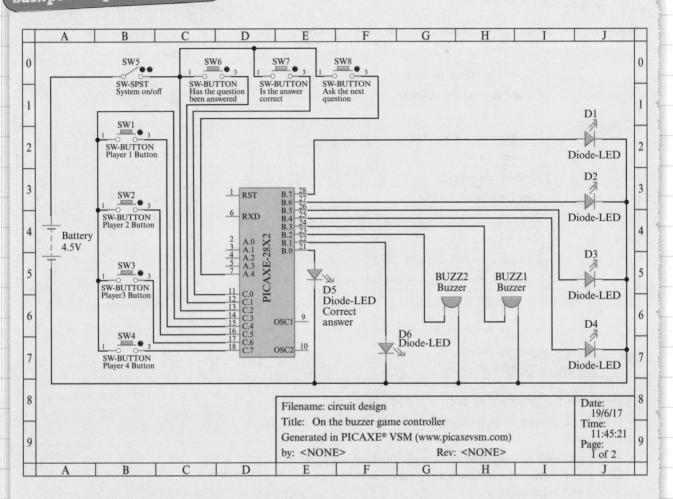

System connections diagram

The diagram is my initial design for the system. SW5 is the on/off control for the whole system. The power connections for the microcontroller are not shown as this makes the diagram clearer. SW1–4 indicate if a player pushes a button, as do LEDs D1–D4. SW6 allows the quiz master to allow the system to proceed once a question has been answered. SW7 and 8 indicate if the answer is right or wrong. The two buzzers indicate when a question has been answered and when the time allowed to answer has elapsed.

In this extract from a response, the diagram makes it clear **which player uses each button**, but does **not** specify **which LED is associated with each player**. Adding this information would **improve** the response.

Now try this

Describe how you would annotate the diagram to link each player with an LED.

Planning a program structure

Here is an example of **initial design proposals** for the **software** components of the project.

Sample response extract

Outline plan for program structure – decision table

		Rules							
Conditions	Player 1 pushes button	✓	✓						
	Player 2 pushes button			✓	✓				
	Player 3 pushes button					✓	✓		
	Player 4 pushes button							✓	✓
	Answer correct	✓		✓		✓		✓	
	Answer incorrect								
Actions	Turn on players' LED	✓	✓	✓	✓	✓	✓	✓	✓
	Sound question answered buzzer	✓	✓	✓	✓	✓	✓	✓	✓
	10 seconds delay	✓	✓	✓	✓	✓	✓	✓	✓
	Turn on correct answer LED	✓		✓		✓		✓	
	Turn on incorrect answer LED		✓		✓		✓		✓

Sample response extract

Outline plan for program structure – pseudo code

- Wait for a player to push their button (using the *If* command).

- When a player has pushed their button, turn on their LED and a buzzer to indicate they have answered (using the *high* command).

- Wait 10 seconds for the answer (using either *pause* or *wait* command).

- The quiz master decides if the player's answer is correct or incorrect (using the *If* command).

- If the answer is correct, turn on the correct LED (using the *high* command).

- If the answer is incorrect, turn on the incorrect LED (using the *high* command).

- Go back to the beginning and start again. (Use either a loop or subroutine and *return* command) .

> Here, the engineer has offered a proposal for both what to do (wait) and how to do it (*If* command). This approach of 'what and how' will make constructing the complete program a more straightforward process.

> Here, the plan offers two possible methods of achieving a delay. The engineer could add comments indicating how they would choose between the two options.

> Here, the engineer has again provided evidence for 'what and how' but has not considered how long the LED should stay on for. The plan could improve by adding the details shown below:
>
> If the answer is correct, turn on the correct LED (using the *high* command).
>
> Wait three seconds (using the *pause/wait* command).
>
> Turn the LED off (using the *low* command).

> This evidence would be improved by considering how to handle **unexpected events**.

Now try this

Use the design brief to describe what sort of unexpected events might occur that this proposed structure does not consider.

System assembly and programming

You will need to **assemble** the **hardware** element of your solution and **write your program**.

> **Links** To revise system assembly, coding and testing, look at page 223.

Sample response extract

```
1     'Program written by Paige Turner   26 June 2017
2     'Easy County School Centre Number 12345
3     'Controlled Assessment task for Unit 6: Microcontroller Systems for Engineers
4     'This is the final version of the program to be submitted as evidence for assessment
5     'file saved as n:/students/year12/TurnerP/Engineering/Unit6/Programs/Final.BAS
6
7     main:
8           let dirsB = 255               'set port b to outputs
9           let dirsC = 0                 'set port c to inputs
```

Remember to include the **full text** of your program in your task booklet, including the **file path** and **line numbering**.

Annotations are essential to explain how your program works.

> **Links** Revise annotating code on page 224.

Sample response extract

```
'Has player 1 pushed their button
'If they have light their LED
'Give them 10 seconds to answer
'Move to next player
```

Choose **sensible names** for your **variables** and **subroutines**. 'Ten_seconds_answer' is much clearer than 'subroutine_3F'.

```
21    if pinC.2=1 then
22          High B.1
23          gosub ten_seconds_answer
24          end if
25
26    if pinC.3=1 then
27          High B.0
28          gosub ten_seconds_answer
29          end if
```

Remember, any **changes** to your code and debugging should be **logged** in a change template like the one on page 230.

> **Links** Revise debugging a program on pages 194, 196 and 199.

Program checklist

Make sure your code:

- is annotated
- is clearly written with sensible variable and subroutine names.

Now try this

1 Look at the code above. Write suitable annotations for this code.
2 Write code that will prevent the buzzers from operating before they are activated by the quiz master.

You can use any language or platform you are comfortable with.

System testing

To complete **system testing and results analysis**, you will need to test your system and make modifications if required.

🔗 **Links** To revise system testing, look at page 224.

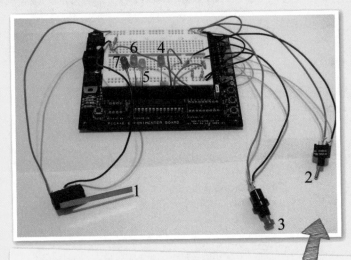

1 Player 1 switch
2 Quiz master correct / incorrect answer
3 Quiz master decision made
4 LED to indicate player 1 has pushed
5 Light to indicate 10 seconds has elapsed (LED has been used instead of buzzer because of noise being annoying during tests.)
6 Green LED to indicate correct answer
7 Red LED to indicate incorrect answer

The program and hardware performs as I expected, so it gives me confidence to expand the system for all four players.

In this extract from a response, the photograph and text **clearly describe** the hardware for the system. Remember to describe how your test was **structured** and the **actual results**. Make sure you **link the judgement of conformity** to the **client brief**.

In this example response, a **problem** and **solution** has been identified as a result of **structured testing**. You need to record solutions and tests in your logs.

I then connected the four player input switches and their associated indicator LEDs. However, pushing switches left to right caused the LEDs to come on right to left. The section of program below caused this.

```
If pinC.O=1 then        'Has player 1 pushed their button
high B.O                 'If they have light their LED
```

By reversing the outputs, such that 1 links with D, 2 links with C, and so on, the display looked better. The code was modified to solve the problem.

```
If pinC.O=1 then        'Has player 1 pushed their button
high B.3                 'If they have light their LED
```

Testing checklist

Make sure that:

• your test results demonstrate that **thorough** and **structured** testing has been carried out
• your judgement of conformity **links test results** to **the client brief**.

Now try this

1 Suggest a method other than a LED to indicate which player has pushed their button.
2 For one of the sample response extracts, write a judgement of conformity that links to the client brief.

Results analysis

Having completed developmental testing, the system is ready to analyse to ensure it performs as required.

Sample response extract

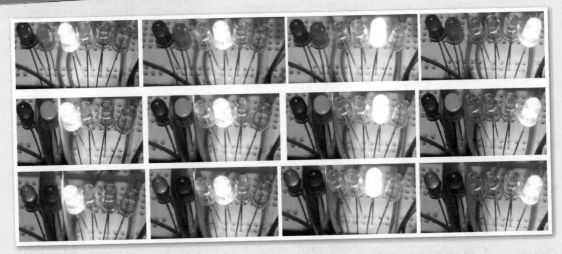

The image shows how the LEDs light up when a player pushes their button. The first row shows from left to right what happens when players 1–4 push their buttons. The second row shows the LEDs when the correct answer is given. The final row shows the LEDs when an incorrect answer is given. The LED indicators all respond as required to player inputs.

> This extract from a response demonstrates that **structured testing** has been carried out on the LED indicators, as all possible **combinations** of results have been **considered** and **evidenced**.

Sample response extract

I thought there would be a potential problem if more than one player pushed their button at exactly the same time. To test this, I connected all four player inputs to a single switch. I operated the switch ten times. There was a reasonably random result of which player's light came on. The results were player 1 = 40%, player 2 = 10%, player 3 = 20%, player 4 = 30%.

The client brief requires 'a system that will tell the quiz master which person was first to push the buzzer'. I think my system will not give any player an advantage over the others due to the particular input button they use.

> In this extract from a sample response, the test results include the simulation of some **unexpected** events. Players pushing their buttons at exactly the same time could be considered an 'unexpected event'.

> You should clearly **link** your **judgement** to the **brief**.

Now try this

Suggest one more possible unexpected event. Describe how you would simulate this and test a system's response.

Recording a system in operation

Once the system is complete you will need to produce an **audio-visual recording** that shows how your **solution** meets the **requirements** of the **design brief**.

> 🔗 **Links** To revise a system in operation, see page 224.

Sample response extract

Plan for content of recording

I need to ensure I cover:

- a functioning system that meets the client brief
- consideration of user experience
- understanding of how the system operates
- understanding of the relationship between hardware and the program
- the use of accurate technical terminology.

> In this extract from a response, the plan for content of the recording meets what is required in the video, taking a sensible approach.

Sample response extract

Detailed plan for content

Section	Content
1	Introduction to me
2	Overview of system hardware
3	Function of player inputs
4	Function of quiz master inputs
5	Function of system outputs
6	What happens when player 1 gives correct answer, with commentary
7	What happens when player 1 gives incorrect answer, with commentary
8	Video of players 2–4 giving correct answers
9	Video of players 2–4 giving incorrect answers
10	While sections 8 and 9 are being recorded visually, audio comments will cover user experience, how the system operates and relationship between hardware and software

> In this extract from a response, the recording has been **broken down** into a **logical sequence**. Make sure you know how much time you have and allow enough time to **show how the system operates** and **provide evidence** of meeting the **client brief**.

Sample response extract

Plan for sections 6 and 7 of the video What happens when player 1 gives an answer

Video content	Audio content
Push button	A micro-switch is used for better user experience
Buzzer sounds and player light comes on	The program keeps the buzzer on for 0.5 seconds
Wait 10 seconds for buzzer to sound again	The player LED will stay on until the QM acts. The colours used for the LED make the user experience more familiar.
Quiz master selects correct / incorrect answer	A toggle switch is used to provide two alternatives with one input.
Quiz master pushes answered button	A push-to-make switch is used so it will be ready to use next time without any user action being needed.
Correct or incorrect LED comes on	Green = correct, Red = incorrect, LED stays on for 1 second
System resets	System is ready for next answer

Now try this

Suggest how the planned content for sections 8 and 9 could be completed within 70 seconds.

> The program will take at least 13.5 seconds for each player to provide an answer. You may want to consider if both these sections need to show evidence for all three players.

Recording a commentary

When planning an audio-visual recording, your commentary should show that you **understand** how your **system operates**. Make sure you use accurate **technical terminology** throughout.

Sample response extract

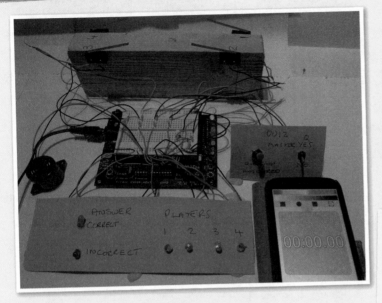

Section 2 audio transcript (overview of system hardware)

'This is my final working system. The player buttons are attached to the block of wood at the back. The micro-switches are labelled 1–4 for players 1–4. The yellow card in the front of the image holds the indicators for when a player answers and the correct / incorrect LEDs. I have included a stopwatch to display timing.'

In this extract from a response, the addition of **labels** to the components ensures the viewer is clear about the **function** of each part. The phone is being used only as a timer. **Balance** the time you spend arranging the hardware with the time available and meeting the requirements.

Sample response extract

Section 7 audio transcript (what happens when player 1 gives incorrect answer)

'As you can hear, 10 seconds after player 1 pushed their button the buzzer came on. When the quiz master pushed the answer button the red incorrect LED came on. This shows this part of my system meets the client brief. The question answered button is a bit fiddly for the quiz master to operate in this prototype, but it should not be too hard to resolve this issue.'

In this extract, the **evidence** and **commentary** has been **linked** back to the **client brief** and has considered **user experience**. Both of these are requirements. However, remember that you need to use **technical terminology** during your recording.

Now try this

This evidence presented does not make use of technical terminology, or the relationship between the hardware and software. Rewrite the commentary for section 7 to include the relationship between hardware and software, using technical terminology.

Answers

Unit 1: Engineering Principles

1. Laws of indices
1 (a) 1   (b) $\frac{1}{4}$ (c) 16   (d) 16
2 $x^{\frac{1}{2}(a+b)}$
3 3

2. Logarithms
1 (a) $\frac{1}{3}$

   (b) $p^3 = 8 \Rightarrow p = 2$

   (c) $\log_4 8 = \frac{3}{2}$

2 (a) $\log_a(5^2) = \log_a 25$

   (b) $\log_a(2 \times 9) = \log_a 18$

   (c) $\log_a(4^3) - \log_a 8 = \log_a\left(\frac{4^3}{8}\right) = \log_a 8$

3. Exponential function
1 128 000
2 4.47 V (2 d.p.)

4. Equations of lines
1 $x + 3y + 9 = 0$
2 $y = \frac{3}{4}x + 5$

5. Simultaneous linear equations
1 $y = -1.37$
   $x = -4.11$
2 $y = 2.10$
   $x = 1.54$

6. Expanding and factorisation
1 $x^2 + 12x - 12x - 16 = 9x^2 - 16$
2 $(3x - 2)(y - 4)$
3 $\dfrac{3V(V_j - V)}{(V_j + V)(V_j - V)} = \dfrac{3V}{(V_j + V)}$

7. Quadratics equations 1
$x = 0.2$ metres

8. Quadratics equations 2
12.8 seconds

9. Radians, arcs and sectors
1 Arc length $= 2.09$ cm (to 2 d.p.)
   Area of sector $= \frac{4}{3}\pi$ cm^{-2} or 4.19 cm/s^2
2 0.45 rad. There are $\frac{2\pi}{0.45}$ sectors in a circle of stock
   material = 13.96. Therefore, 13 complete sectors could be
   formed from one complete circle.

10. Trigonometric ratios and graphs
1

θ (°)	θ (radians)	sine θ	cosine θ	tan θ
0	0	0	1	0
30	$\pi/6$	0.5	0.866 or $\frac{\sqrt{3}}{2}$	0.577
45	$\frac{\pi}{4}$	0.707	0.707	1
60	$\frac{\pi}{3}$	0.866 or $\frac{\sqrt{3}}{2}$	0.5	1.732
90	$\frac{\pi}{2}$	1	0	$-\infty$
135	$\frac{3\pi}{4}$	0.707	-0.707	-1
180	π	0	-1	0
270	$\frac{3\pi}{2}$	-1	0	$-\infty$
360	2π	0	1	0

2 10 cm (i.e. it is an isosceles triangle with two equal sides and
   two equal angles)

11. Cosine rule
(a) $PR = 18.9$ cm (3 s.f.)
(b) the size of $\angle PQR = 1.15$ rad (3 s.f.)

12. Sine rule
11.3 cm

13. Vector addition
Magnitude = 127 N
Direction = 14° (above the horizontal)
Sense = +ve (where left to right is taken as the positive direction)

14. Surface area and volume
1 18.47 m^3 (2 d.p.)
2 volume is $2^3 = 8$ times bigger

15. Systems of forces

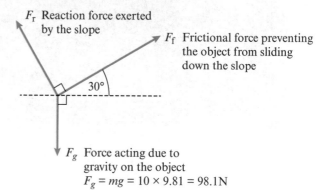

F_r Reaction force exerted by the slope
F_f Frictional force preventing the object from sliding down the slope
30°
F_g Force acting due to gravity on the object
$F_g = mg = 10 \times 9.81 = 98.1$ N

16. Resolving forces
Magnitude = 24.29 N
Direction = 25.17° (anticlockwise from the vertical)

17. Moments and equilibrium
The plate is not in static equilibrium.
$\Sigma F_v = -7.22$ N (where vertically up is +ve)
$\Sigma F_h = 15.33$ N (where left to right is +ve)
$\Sigma M = 22.4$ N m (where clockwise is +ve)

18. Simply supported beams
$\Sigma F_v = 12 + 62 - 45.33 - 28.67 = 0$ (where vertically up is +ve)
$\Sigma F_h = 0$
$\Sigma M = 0$

19. Direct loading
8 333 333 Pa or 8.33 MPa

20. Shear loading
24.32 MPa

21. Velocity, displacement and acceleration
(a) $a = -1.75$ m/s^2
(b) 150 m

22. Applying the SUVAT equations
(a) The boat decelerates at 0.4 m s^{-2}.
(b) $v = 4$ m/s

23. Force, friction and torque
1 329.08 N
2 17.5 N m

24. Work and power

1 2000 N
2 5.25 kW (2 d.p.)

25. Energy

1 377.69 MJ
2 $v = 12.13$ m/s

26. Newton's laws of motion, momentum and energy

1 39 m/s
2 $v = 7.6$ m/s

27. Angular parameters

1 0.534×10^6 W
2 1263.31 J (2 d.p.)

28. Mechanical power transmission

1 84.2% (3 s.f.)
2 $VR = 30.67$
   $MA = 16.3$
   Efficiency = 53.1% (3 s.f.)

29. Submerged surfaces

1 90 497 N or 90.50 kN
2 2 m

30. Immersed bodies

740 kg/m^3

31. Fluid flow in tapering pipes

31.25 m/s

32. Heat transfer parameters and thermal conductivity

17 280 W

33. Heat transfer processes

Conduction requires contact between atoms or molecules to transfer heat in the form of kinetic energy. The vacuum established in the walls of the flask means that there is no medium through which heat can be transferred by conduction.

Convection requires a fluid in which to establish a convection current to transfer heat energy. Again, the vacuum established in the walls of the flask means that there is no fluid medium through which heat can be transferred by convection.

Radiation transfers heat energy through electromagnetic radiation, which is able to travel through a vacuum. However, the internal walls of the vacuum flask have a highly reflective coating to reflect back as much of this electromagnetic radiation as possible.

34. Linear expansivity and phases of matter

100.15 mm

35. Specific heat capacity, latent and sensible heat

2679 kJ

36. Heat pump performance ratios

2.22 kW

37. Enthalpy and entropy

Internal energy (U) is the energy present in an amount of a substance and comprises the potential energy stored in atomic and molecular bonds and the kinetic energy present due to the random vibrations and movements of atoms and molecules. However, this is only part of the total energy, or enthalpy, contained in a thermodynamic system. Enthalpy includes both internal energy (U) and the work energy (pV), which can be thought of as the work that the substance has done on its surroundings.

38. Thermal efficiency of heat engines

0.19 or 19%

39. Thermodynamic process parameters

294 357 Pa

40. Gas laws

208.15 k Pa

41. Current flow

1 $I = 1.5 \times 10^{-3}$ A
2 2 days

42. Coulomb's law and electrostatic force

1 $F = -5.39 \times 10^2$ N (2 d.p.)
2 Two identical charges of 1.15×10^{-6} coulombs, separated by 20 mm, produce a repulsive force of 30 N.

43. Resistance, conductance and temperature

1 3.33×10^{-3} °C^{-1} (3 s.f.)
2 0.01 siemen

44. Types of resistor

1 Any of the following make this type of resistor unsuitable for high specification uses:
   • they have poor stability over time
   • the resistance value can change if voltages close to the maximum rating are applied
   • they are susceptible to changes in humidity
   • the nominal resistance values may change during soldering.
2 5.0 MΩ with a tolerance of ±5%

45. Field strength

1 10×10^3 N/C
2 $E = \frac{300}{0.04} = 7.5 \times 10^3$ V/m

46. Capacitance

17.70×10^{-12} F or 17.70 pF

47. Capacitors – non-polarised

Each terminal of the capacitor is connected to a conducting sheet; e.g. metal foil. The dielectric must be an insulator to prevent leakage of charge from one plate to the other.

48. Capacitors – polarised

1 Correct terminal connection is required to avoid breaking down the insulating oxide layer. It may be indicated by any of the following: painted arrow, with + or – pointing to relevant pin, a red dot next to positive pin or one lead longer than the other
2 5 V

49. Ohm's law, power and efficiency 1

1   For the graph see page 56.

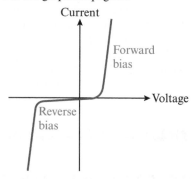

An ohmic resistance produces a straight line graph when p.d. V is plotted against current I. The line passes through the origin and its gradient is the resistance R. The diode plot of V against I is not a straight line that passes through the origin and therefore does not conform to Ohm's law so is not an ohmic device.

2   $I = 13.63\,\text{A}$

50. Ohm's law, power and efficiency 2

1   $P = IV$ For power to remain constant, if voltage increases by 4 then the current would reduce to a quarter of its original value.
2   Efficiency = 60%

51. Kirchoff's voltage and current laws

$5\,\Omega$

52. Capacitors – charging and energy

$0.02\,\text{C}$ (2 d.p.)

Note: this is the maximum charge. Capacitors charge / discharge exponentially so the capacitor would need to charge for an infinitely long period of time to charge it to maximum capacity. See pages 54 and 55 for a more practical approach.

53. Capacitors – networks

1   Capacitors in parallel will produce the greatest capacitance
2   $23.96\,\mu\text{F}$ (to 2 d.p.)

54. Capacitors in circuits – the time constant

(a) $0.0264\,\text{C}$ (b) $220\,\text{s}$

55. Capacitors in circuits – *RC* transients

1   $37.3 \times 10^{-6}\,\text{F}$ or $37.3\,\mu\text{F}$
2   A capacitor is normally considered fully charged after $5\,\tau$, which would give a total charge time of $5 \times 5 = 25$ seconds.

56. Diodes – bias and applications

The diodes, AC supply and DC load are arranged as shown.

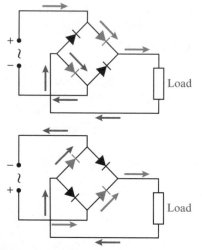

During the first half cycle the diodes shown red allow current to flow around the circuit. The two other diodes are in reverse bias mode and therefore no current flows through them. During the second half of the cycle the other two diodes allow current to flow through the load, whilst the opposite diodes are now in reverse bias mode.

57. DC power sources

$11.37\,\text{V}$

58. Resistors in series or parallel

$694.39\,\Omega$

59. Resistors in series and parallel combinations

$23.89\,\Omega$

60. Resistors and diodes in series

$1383\,\Omega$

61. Capacitors in series or parallel

$2.2 \times 10^{-4}\,\text{C}$
$4.4 \times 10^{-4}\,\text{C}$
$5.5 \times 10^{-4}\,\text{C}$

62. Capacitors in series and parallel combinations

$1.13 \times 10^{-4}\,\text{C}$

63. Magnetism and magnetic fields

$12\,000\,\text{A/m}$

64. Permeability

$11.06\,\text{T}$

65. *B/H* curves, loops and hysteresis

A significant cause of energy loss that limits the efficiency of transformers are hysteresis losses caused by the magnetisation/demagnetisation cycle. The amount of energy lost as heat is proportional to the area inside the hysteresis loop when plotted as a *B/H* curve. Materials with an extremely narrow hysteresis loop are chosen for use in transformers to minimise these losses.

66. Reluctance and magnetic screening

$83\,766\,\text{1/H}$

67. Electromagnetic induction

$0.12\,\text{m}$

68. DC motors

The commutator feeds electric current to each of the motor coils in turn as it rotates as part of the armature. The stationary spring-loaded carbon brushes maintain contact with the commutator as it rotates.

The armature comprises the drive shaft, commutator and coils wrapped around a laminated iron core. The armature is the driven component of the motor and rotates in bearings to reduce friction losses.

69. Electric generators

Increases in the following will increase the emf generated:
• the number of turns in the coil
• the speed of rotation
• the strength of the magnetic field in which the coil rotates.

70. Inductors and self-inductance

$0.099\,\text{J}$

71. Transformers and mutual inductance

$95.8\,\text{V}$

72. AC waveforms

62.5 Hz

73. Single phase AC parameters

155.56 V

74. Analysing AC voltages using phasors

$v = 291\sin(100\omega t + 0.351)$

75. Reactance and impedance

(a) $377\,\Omega$
(b) $610\,\text{mA}$

76. Rectification

The 240 V AC supply would first be reduced to 5 V AC using a step down transformer. This must then be changed to DC using a full wave rectifier circuit with an output smoothing capacitor. The bridge rectifier circuit with smoothed output shown below would be suitable for this application:

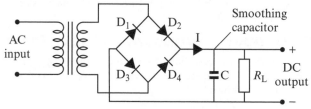

The smoothed 5 V output voltage would look like this:

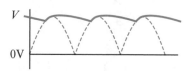

77. Your Unit 1 exam

Your notes on the Unit 1 exam, always referring to the latest Sample Assessment Material on the Pearson website for an indication of assessment details.

78. Showing your working

165.5 kN

79. State and describe questions

A slip ring is connected to each end of the rotating coil inside the alternator. When the coil rotates and cuts through lines of magnetic flux an emf is generated in the coil and so between the slip rings. Stationary spring-loaded carbon brushes maintain contact with each slip ring as it rotates. This allows the emf generated to be transferred to the output terminals of the alternator.

80. Explain questions

A hair dryer operates on the principle of forced convection, where a fan is used to move air over a heated element. Heat energy is transferred to the air by conduction as it passes over the element. A continuous supply of cool air is heated in this way. This process differs from conventional convection because the movement of air is not caused by the variation in density caused by local heating and expansion but is forced using a fan.

81. Find questions

1.17 m

82. Calculate questions 1

85.49 m

83. Calculate questions 2

1716 N

84. Solve questions

0.199

85. Draw questions

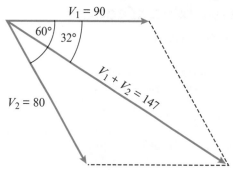

By measurement from diagram:
Magnitude of resultant phasor = 147
Phase angle (below the horizontal) = 32°

86. Label questions

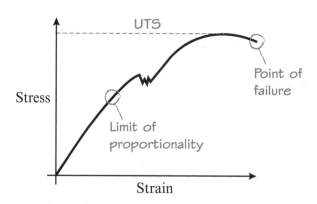

87. Identify questions

1 thyristor
2 variable capacitor
3 Zener diode

88. Synoptic question 1

0.993 or 99.3%

89. Synoptic question 2

0.213 or 21.3%

90. Using the formulae booklet

1   $v_c = Ve^{(-t/\tau)}$ is across the capacitor
   v_c is initial voltage across capacitor
   t is time
   τ is time constant of the circuit

2   $F = \dfrac{q_1 q_2}{4\pi\varepsilon_o r^2}$

   F is the force between two charged particles
   q_1 is charge of first particle
   q_2 is charge on second particle
   ε_o is permittivity of free space
   r is the distance between the particles

3   $F = \rho g A x$
   F is hydrostatic thrust
   ρ is fluid density
   g is acceleration due to gravity
   A is area
   x is the distance from the free surface to centroid of the submerged plane surface (the point at which average pressure acts)

Unit 3: Engineering Product Design and Manufacture

96. Design triggers 1

Example answer:
1   Market pull – led by ever-changing fashions within the industry and consumers copying the designs within the large exclusive fashion houses
2   Technology push – due to the very competitive nature and tight regulations of F1 racing, the teams rely on research and development into new technologies to gain an advantage over their rivals.
3   Combination of market pull and/or technology push – mobile phone manufacturers are heavily influenced by market pull and use product reviews, feedback and market research to ensure their products meet basic customer expectations of functionality and aesthetics. However, technology push also has an important role in providing unique selling points that can give manufacturers a competitive advantage over their rivals; e.g. the inclusion of digital cameras in mobile phones and, more recently, the introduction of mobile virtual personal assistants were both the result of technology push.

97. Design triggers 2

Answers might include:
1   As small children tend to put toys in their mouths during play, all the materials used in the manufacture and packaging of the product should be non-toxic.
2   As toys are often dropped or thrown during play, any corners or edges should be rounded off to reduce the risk of causing injury should they come into contact with another child.
3   Exposed mechanisms or moving parts should be avoided, as they may present a trapping hazard to children's hands or fingers during play.

98. Reducing energy

Answers might include:
• Using a virtual design environment reduces the need for producing physical prototypes and mock ups, which are energy intensive, time consuming and expensive to produce.
• Using a virtual design environment reduces the likelihood of costly design errors by enabling thorough testing of each virtual component in situ prior to manufacture. Any problems, such as clashes with other components, can be addressed quickly with minimum additional expenditure of energy, time or expense.

99. Hybrids and energy recovery systems

Answers might include:
• Typically, hybrid vehicles achieve better fuel economy than a conventional vehicle that uses just an internal combustion engine.
• Hybrid vehicles produce lower levels of harmful emissions than a conventional vehicle that uses just an internal combustion engine.

100. Sustainability and cost over product life cycle

Answers might include:
1   Minimising the volume of the product by supplying it as a flat pack minimises storage, transport and packaging costs.
2   Minimising packaging reduces the amount of packaging materials used and so increases sustainability.

101. High-value manufacturing and designing out risk

Answers might include any two of the following:
1   • insulated sleeves on part of live and neutral pins to prevent accidental contact with fingers as plugs are inserted
    • an integral fuse providing overload protection and disconnecting the electrical current when the maximum safe current is exceeded
    • a cable clamp prevents wires from being pulled out of the plug and exposing live conductors
    • during insertion into a socket, the slightly longer earth pin opens shutters that usually cover the live and neutral connections so that even small fingers can't touch live parts
    • the plug has to be disconnected from the mains in order to access the screw that secures the cover. This prevents removal of the cover whilst the plug is connected and prevents the exposure of live parts inside the plug.
2   A 3-pin plug has the characteristics of a commodity product. It is cheap to buy and is manufactured in high numbers (low value, high volume).

102. Systems, equipment and interfaces

Individual response, depending on chosen field of study.
For example, the system used to operate anti-lock brakes on a motor vehicle is shown below.

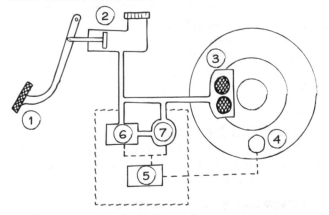

1   The driver presses the brake pedal.
2   A piston pushes brake fluid out of the master cylinder.
3   Pressurised brake fluid activates the brake callipers, which force the brake pads into contact with the brake disc, slowing the rotation of the wheel.
4   A sensor monitors the rotation of the wheel. If this detects that the wheel is about to lock up then a signal is sent to the ABS system.
5   A microcontroller in the ABS system receives a signal from the sensor that the wheel is about to lock up. This triggers a response from the ABS system.
6   The microcontroller operates a valve that redirects brake fluid into an accumulator, relieving the pressure in the fluid and releasing the brakes.
7   Almost immediately after this, the microcontroller closes the valve into the accumulator and pumps the fluid out, re-pressurising the system and re-applying the brakes.
8   Steps 6 and 7 repeat several times a second until the sensor detects that the risk of locking up the wheel has passed.

103. System compromises

Individual response, depending on chosen field of study.
For example, on the ground many aircraft are de-iced by spraying them with de-icing chemicals. In the air this becomes impractical. Instead, waste heat from the engines is redirected into the wings and used to maintain a leading edge temperature above $0\,°C$, preventing the formation of ice whilst the aircraft is in flight.

104. Equipment specifications and cost effectiveness

Individual responses. For example, using a systems approach to design allows manufacturers in the automotive industry to carry out concurrent product development in areas such as: powertrain (engine and gearbox), chassis and structural components, body and interior styling. This work is often done in global centres of excellence, using specialist teams of engineers and designers. Only when all these separate elements are integrated into a complete system design is a new model born.

Motor vehicle manufacturing takes a similar approach, with major systems such as engines and gearboxes being manufactured in specialist plants situated across the globe. These are only brought together when they arrive in the assembly plant that manufactures finished vehicles, so great care must be taken to ensure they will fit and function seamlessly together.

105. Mechanical properties

Example answer:
- Hardness. High hardness will allow the drill cutting edge to slice through less hard materials, such as mild steel.
- Toughness. High toughness will allow the drill to resist brittle fracture if subjected to impact or shock loading whilst drilling.

106. Physical and thermal properties

Example answer:
- Physical property – high melting point to prevent the material from melting when exposed to the high temperatures encountered during re-entry.
- Thermal property – low thermal conductivity to limit the transmission of heat and insulate the main body of the capsule from the high temperatures encountered during re-entry.

107. Electrical and magnetic properties

Example answer:
Reluctance. The presence of a magnetic field will cause magnetic flux to flow around a path of least magnetic reluctance. By using a shielding material with low reluctance, magnetic flux will flow through it in preference to high reluctance air and so can be redirected away from and around the area being shielded.

108. Advanced materials

Example answer:
A super elastic smart alloy such as nitinol could be used in spectacle frames to eliminate the need for mechanical hinges otherwise required to allow the arms to be folded. An additional benefit is that if the frames are accidentally sat or even stood on they will suffer no permanent damage and simply spring back into shape.

109. Surface treatments and coatings

Example answer:
Maximum corrosion protection would be provided by hot dip galvanising the chassis and then applying a further protective layer of specialist chassis paint.

110. Lubrication

Example answer:
- Many of the bearings in an engine rely on lubrication provided by pressurised engine oil to prevent metal on metal contact that would otherwise cause rapid wear and premature failure.
- Circulating engine oil prevents bearings and other moving parts from overheating by carrying away heat from them.
- Circulating engine oil removes small particles of metallic debris away from moving parts where they might otherwise cause damage. The debris is then removed from the circulating oil by the oil filter.

111. Modes of failure of materials

Example answer:
Any metal, including the alloys used in turbine blades, that is required to operate at high stress levels and elevated temperatures is likely to be affected by creep. A major contributor to the gradual plastic deformation that characterises creep is gradual movement between the material crystal grains where they meet at grain boundaries. To help eliminate this effect, turbine blade manufacturers have developed manufacturing methods to cast creep-resistant turbine blades in a single crystal so that no grain boundaries are present.

112. Mechanical motion

Answers might include:
Linear motion: hydraulic ram, linear actuator, solenoid.
Rotary motion: prop shaft, gear, sprocket, pulley.
Reciprocating motion: engine piston, reciprocating saw.
Oscillation: pendulums used to measure time, tuning forks used to measure frequency, most forms of unwanted mechanical vibration are caused by oscillation.

113. Mechanical linkages

Individual responses. For example, the linkage that operates a windscreen wiper.

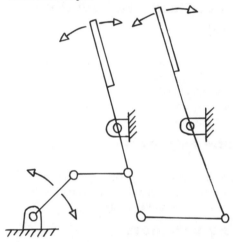

114. Power sources

Responses might include:
1. Internal combustion engine: providing the mechanical power to drive a mobile generator.
2. Lead acid battery: providing high current electrical power to drive a fork lift truck.
3. Photovoltaic cells: providing electrical power to operate a pocket calculator.

115. Controlling power transmission

Individual responses. For example:
Tensile testing machine. A tensile testing machine is designed to maintain a constant strain rate in a material sample whilst monitoring the applied tensile force.
Sensors. Positional sensors monitor the position and movement of the cross heads on which the jaws are mounted as they move apart. Load sensors monitor the force being exerted on the material sample during the test.
Controllers. The signals from the sensors are processed by a microprocessor (often interfaced with a PC). This not only records all the test data, and automates the generation of graphs and standard test reports, but also controls the actuators used to apply the tensile force during the test.

Actuators. A hydraulic cylinder or ram is commonly used to provide the necessary force to move the cross heads apart and exert the required load on the sample during testing.

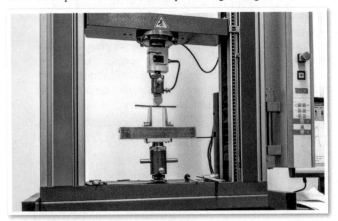

116. Processing metals

Individual responses. For example:
The most cost-effective method would be to produce the cross sectional profile of the bracket in long lengths, using extrusion. Individual brackets would then be cut and drilled.

117. Powder metallurgy and additive manufacturing

Answers might include:
Components produced by power metallurgy have no sprues, flash or joint lines that have to be removed in secondary operations. Components can be manufactured using a combination of metallic and non-metallic materials by controlling the composition of the powder mix.

118. Joining and assembly

Example answer:
Brazing is usually used in this type of application. Brazing is carried out at a lower temperature than welding, so avoids the effects of heat distortion on thin-walled components.

119. Processing polymers

Example answer:
Injection moulding is usually used in this type of application. Injection moulding is suitable for high volume thermoplastic components with complex shapes, good surface finish and good dimensional accuracy.

120. Processing ceramics

Example answer:
Selective laser sintering (SLS) is usually used in this type of application. SLS will allow the cost-effective production of a single prototype component directly from a computer model without the time and expense required to manufacture a mould.

121. Processing fibre-reinforced composites

Example answer:
GFRP is suitable for making a garage door as:
- It can provide adequate strength, rigidity and toughness for this non-load bearing application.
- The glass fibre and polyester resin materials are less expensive than traditional alternatives for garage doors such as hardwood.
- GFRP will not corrode or rot and has a long service life.
- A coloured gel coat or overpainting can provide a range of colour options.

The spray-up process makes it quick and easy to manufacture large components, such as a garage door, using a simple and inexpensive open mould; it allows the door to be manufactured in one piece, which avoids the cost of additional joining and assembly processes.

122. Effects of processing

Answers might include:
To reinforce the component in a particular direction, re-orientate single fibres or fibre tows along the line of action of the applied load. This provides additional strength in that specific direction. A less efficient option would be to increase the ratio of fibre to resin used in the composite. Using a greater proportion of woven matting will provide an overall increase in strength in all directions.
In some situations, laminating by adding additional layers of CFRP is a simple way to strengthen components. However, this is unsuitable here as it would increase the size and weight of the component.

123. Scales of production

Individual responses. For example:
One-off: a prototype component.
Small batch: 25 brackets
Large batch: 800 gears
Mass production: cars (assembly); crankshafts (manufacturing)
Continuous production: machine screws

124. Customers

(a) Individual responses. For example, an on-site aluminium can manufacturing plant supplies the main bottling facility at a soft drinks manufacturer (internal customer) and is asked to update the branding printed on the cans.
(b) Individual responses. For example, a machine tool manufacturer supplies a CNC milling machine to a company making bicycles (external customer) and is asked to provide specialised work-holding clamps to enable the machining of a particular component.

125. Product and service requirements

Individual responses. For example:
Consider the product requirements for a pocket calculator:
- Performance specifications – processing speed sufficient to provide instantaneous responses once a calculation has been entered.
- Operating standards – calculations entered into the calculator must be processed in the standard mathematical order defined by BIDMAS (brackets, indices, division, multiplication, addition, subtraction).
- Manufacturing quantity – high volume (100 000+).
- Reliability – user expectation for reliability is high. The calculator must not make errors in calculations and remain fully functional for 10 years+ under normal conditions.
- Product support – operating instructions will be required.
- Life cycle – users must have access to replace product battery in order to extend the life of the product.
- Usability – keys should be laid out in a conventional and logical sequence, with clear and easy-to-understand labels.
- Anthropometrics and ergonomics – the size of the buttons and the distance between them must be sufficient to allow operation of a single button with an adult index finger without interfering with or operating surrounding buttons.

126. Product design specification (PDS) 1

Individual responses. For example:
Consider the PDS for a smartphone.
1 Cost – If based on a retail selling price of £500 then perhaps £100 should be the target manufacturing cost.
2 Quantity – high volume (100 000+)
3 Maintenance – other than the need for battery charging the product should be maintenance free.
4 Finish – matt self-finish (unpainted) exposed surround, high gloss black back and front screen surround.
5 Materials – lightweight but tough and strong aluminium frame, scratch and shatter-resistant glass back and screen.
6 Weight – target weight <150 g.

7   Aesthetics – to match brand image for simple, elegant styling.
8   Sustainability – use of clips and fixing screws instead of adhesives allow straightforward disassembly for recycling or repair and reuse.

127. Product design specification (PDS) 2

Individual responses. For example:
To continue the example from page 126, consider the PDS for a smartphone.

1   Reliability – user expectation of reliability is high. A life of 5+ years without any reliability issues is expected.
2   Safety – user expectation of safety is high. Recent issues with one model of smartphone in which batteries overheated causing fires led to that model being withdrawn completely at a cost of billions to the company involved.
3   Testing – rigorous product testing will be required prior to product launch. That will include hardware functional testing, including battery charging, battery life, reception and transmission capabilities. Also, software testing is extremely important to ensure that this runs smoothly and functions correctly without bugs.
4   Ergonomics / anthropometrics – physical size must sit comfortably in the hand.
5   Usability – user interface should be intuitive so that users can navigate the functions and different features of the phone easily without having to make repeated reference to operating instructions.

128. Commercial protection

(a) Copyright
(b) Design registration
(c) Trademark

129. Legislation and standards

Individual response. For example:
Having to make changes to meet new regulations will provide the manufacturer with an opportunity to make any additional improvements or refinements necessary to revamp the design and functionality of the product and/or reduce manufacturing costs. If they can be first to market with a product that complies with the new standard, then this might also provide them with a commercial opportunity to increase sales of the updated model.

130. Environmental and safety constraints

Individual responses. For example:
Continuing the example considering an aluminium drinks can.
Raw material processing – rolling the large ingots of aluminium (that are the product of the ore extraction and refinement) into thin sheets is an energy intensive process reliant on fossil fuels to provide the electricity to run the large and powerful equipment involved.
Manufacturing – round blanks for each can are pierced from a roll of thin aluminium sheet. The surrounding waste material is recycled, minimising waste. These blanks are then shaped into cans by cold forming in a series of cupping and roll-forming operations. Heating is not required during these processes, minimising energy consumption, and no waste or potentially harmful emissions or byproducts are produced.
Packaging – cans are themselves items of packaging. Cans are fastened together using 4-pack or 6-pack polymer rings. Bulk packs are further secured on cardboard trays with polymer wraps. Polymer-based packaging is derived from oil, and thin films and bands are not widely recycled and often end up in landfill.
Distribution/Transport – empty cans have low weight but high volume and are easily damaged in transport. For this reason, can manufacturing plants are usually sited within or next to the bottling plants in which they are used. This also means that any environmental impact from transportation is avoided.

Use/Reuse – cans are designed to be disposable. Once the ring pull is opened cans cannot be re-sealed.
End of life – cans are designed to be disposable and have a very short life.
Recycling – although some aluminium cans do end up in landfill, the majority are recycled after use. Unlike some polymers, there is no limit on the number of times aluminium can be recycled. Recycling uses only a fraction of energy used extracting and refining aluminium from its ore and so makes good environmental and commercial sense.

131. Security constraints

Individual response. For example:
Consider the security features in your mobile phone.
Mobile phones contain a large of amount of personal and financial information about us that could be misused if it fell into the wrong hands. The software that operates smartphones has to be resistant to malicious attempts to access the contents of your phone, clone it, intercept messages and email, or listen in to your conversations. All smartphones have an access PIN or password that must be used to gain access to it, some have fingerprint recognition systems and others allow remote access, should they be lost or stolen, to either track their whereabouts or wipe their contents.

132. Marketing

Individual responses. For example:
Consider the features of a pocket calculator.
USP: It's specially designed for 16 years+ students and contains all the functions required in my A level maths course.
Competition: It's manufactured by the brand leader in pocket calculators, which has an established reputation that I trust.
Lifetime: At least five years, although it will probably last much longer.

133. Form and functionality

Individual response. For example:
Any two of the following functional requirements: is freestanding in the upright position, has an internal chamber for collecting shavings that can be opened easily for emptying, is able to sharpen pencils easily and effectively.
Any two of the following form requirements: it resembles a stylised beaver, has a highly polished decorative finish, is playful and humorous (beavers eat wood), the pencil is inserted into the beaver's mouth.

134. Product performance

Individual response. For example:
Any two of the following characteristics:
Low conductivity – the material must be a poor electrical conductor in order to insulate the electrical connections from each other and from the user of the plug.
High toughness – cracks or breakages caused by dropping or other impact might expose the live conductors inside the plug, which would present a serious hazard to the user.
Suitability for moulding processes – the complex curved shape of the case can be manufactured most efficiently using a moulding process.
Stiffness – the case must be stiff in order to retain the three pins in the correct position and orientation during use.
Fire/heat resistance – overloading or an electrical fault can cause overheating. The plug body must retain its stiffness and other properties at elevated temperatures. In extreme circumstances electrical faults can cause fires. The material used in the plug body must not be flammable or support combustion.

135. Manufacturing processes and requirements

Example answer:
Die casting. Manufacturing requirements will include the need for investment in complex tooling and equipment, the manufacturing process is highly automated and requires low-skilled labour, allowing economical manufacturing only in high volume.

136. Manufacturing needs

Milling
Turning

137. Generating design ideas

Individual responses. For example:
Responses might include sketches and annotation considering the following temporary solutions for carrying loads on the vehicle roof: inflatable pads, foam pads, straps, elastic bungees, suction pads, hook and loop tape, netting, slings, detachable frames, corrugated card supports etc.

138. Development

Individual responses. Suitable areas for investigation might be:
Material for the frame (to enable possible weight reduction), its type and sizing options
Saddle type, size, and comfort
Position and operation of brake levers

139. Design information

Resistivity of nichrome wire is in the region 1–$1.5 \times 10^{-6}\,\Omega\,m$. You could have found this information from a number of sources including:
Mechanical Engineers Pocket Book
Manufacturer's datasheet
Online material database

140. Freehand sketching, diagrams, technical drawings

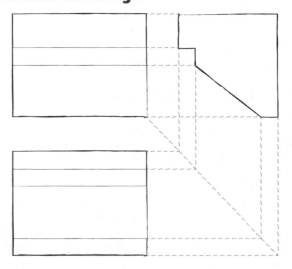

141. Graphical techniques

Individual responses.
The best technique for enabling easy comparison of the data is the pie chart.

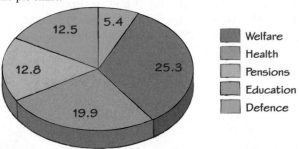

142. Written communication

Individual responses. A suitable response might be:
The rear section of the plug housing suffered considerable damage caused by multiple brittle fractures when subjected to an impact with a hard floor. The material was too brittle and lacked the toughness required in this application. The materials should be replaced with an alternative with improved toughness to make it more impact resistant.

143. Design documentation

Example answer:
1 If the component was not on the BOM then they would not have been ordered from the supplier prior to the start of manufacture / assembly. When the error was noticed on the production line there would be significant delay and disruption, as the completion of the products would not be possible. Any partially completed work would have to be stood off in a holding area until the components required to finish it were ordered and delivered. Customer delivery promise dates would not be met, the reputation of the company would suffer and future orders might be put at risk.
2 The component would not fit as required during assembly. The components would have to be re-worked or scrapped and made again. As in the previous example, there would be significant delay and disruption. Customer delivery promise dates would not be met.

144. Iterative development

Individual responses. Suitable examples would be electronic tablets, laptop computers, games consoles, software apps and operating systems.

145. Statistical data 1

Discrete data: mean = 123.75
Continuous data: mean = 52.26

146. Statistical data 2

Standard deviation is the square root of the variance = 1.799.

147. Data handling and graphs 1

Frequency distribution table

Number	Frequency
1	1
2	2
3	0
4	1
5	2
6	3
7	1
8	1
9	3

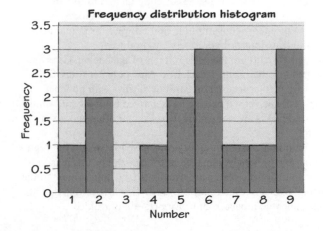

148. Data handling and graphs 2

Bar chart displaying discrete data on machine breakdowns per month between January and June.

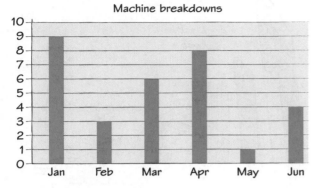

149. Frequency distributions

99.7% of the values in a normal distribution lie between ± 3 standard deviations of the mean.

So the lower limit = 48.8 − (3 × 0.17) = 48.29

Upper limit = 48.8 + (3 × 0.17) = 49.31

150. Validating the design

1

Evaluation criteria from PDS	Weighting	Score (out of 10)	Weighted score
Performance	0.1	9	0.9
Comfort	0.4	4	1.6
Aesthetics	0.2	6	1.2
Weight	0.2	9	1.8
Cost	0.1	8	0.8
		TOTAL (out of 10)	6.3

2 Comfort. This has the highest weighting but the lowest score.

151. Benefits and opportunities

Example answer:

Motherboard circuit redesign to accommodate new pin configuration.

Power supply that must supply more electrical power.

Cooling system that must extract more heat.

152. Further modifications

Example answer:

Due to the draft angle, the castings do not have parallel sides. Work holding in a standard machine vice (that has jaws that move in parallel) might prove problematic.

153. Your Unit 3 set task

Your notes on the Unit 3 set task, always referring to the latest Sample Assessment Material on the Pearson website for an indication of assessment details.

154. Reading a brief

Individual responses from a wide range of relevant information sources that might include, for example: manufacturer catalogues, industry magazine articles, product data sheets, product videos, engineering textbooks.

155. Conducting research

Individual responses.

Manufacturing methods for metal brackets might include:

- Die casting
- Permanent mould casting
- Sand casting
- Forging
- Fabrication (welded, brazed or using fixings)
- Investment casting
- Powder metallurgy
- Stock removal from a solid billet.

If researching stock removal from a solid billet of material (e.g. mild steel, high carbon steel, stainless steel, aluminium) to manufacture brackets, for example, the following factors would be useful to note:

- Although complex geometry is possible, there are limits to what can be machined by conventional methods (e.g. turning, milling, drilling).
- Automation can be used to speed up manufacturing and ensure consistency (e.g. CNC).
- Waste material (e.g. swarf) is easy to recycle.
- It is not necessary to invest in custom tooling, moulds or patterns.
- Modifications to the design can be made easily.
- Heat treatment can be used after machining to improve the mechanical properties of certain materials.

156. Making notes

Research sheet could contain information relevant to the following topics:

Function of the product, Manufacturing methods and processes, Costs, Materials and material properties, Health and Safety issues, Sustainability and environmental factors, Relevant numerical data, Aesthetics, Ergonomics, Anthropometrics, Applied finish, Advantages or disadvantages of the design. There may also be other information needed that is very specific to the topic (can crusher) being researched.

157. Reading further information

Example answer:

1 The brackets are fracturing in service.

2 The average life expectancy of bracket A is 24 years.

3 Each mounting hole is 12 mm in diameter.

158. Interpreting an engineering drawing

1 Overall length 95 + 8 + 100 + 10 + 105 = 318 mm

2 Overall width 94 + 12 + 94 = 200 mm

3 Overall height 105 + 10 = 115 mm

4 Diameter of cable hole = 2 × radius = 2 × 32.5 = 65 mm

159. Analysing data

A 24 years, B 20.6 years C 15.9 years D 11.8 years E 8.3 years

160. Creating a time plan

The time needed for each of the stages will be in response to information on timing and marks from the Pearson website.

161. Recording changes and action points

1 Example answer:

Carbon fibre may be an improvement because it is resistant to corrosion. It may be more expensive and difficult to manufacture. It may be lighter but still strong enough. Carbon fibre may be cost effective in the long term.

2 Suggested actions needed:

Establish cost of carbon fibre compared to steel. Compare manufacturing methods with those for steel. Research life span and strength of carbon fibre compared to steel.

162. Interpreting a brief

Interpretations of the brief should be similar to the examples below:

Regulations:
- The product must comply with The Sale of Goods Act 1979 if it is to be retailed in the UK. It must be fit for purpose.
- The product must comply with The General Product Safety Regulations 2005 if it is to be retailed in the UK. The manufacturer has a general duty to ensure it is safe to use.

Sustainability:
- The product must be packaged only in biodegradable material.
- The product must be manufactured from materials that are 100% recyclable.

Requirements:
- The dispenser must be suitable for the standard 24 mm × 66 mm roll of sticky tape.
- The product must not scratch or deface the surface it is being used on.

Constraints:
- The dispenser must be no bigger than 150 × 100 × 100 mm.
- The product should retail at less than £10.

Health and safety:
- It must not be possible for the user to injure themselves on the tape-cutting element of the product.
- The product must be supplied with clear, multilingual instructions, explaining procedures for correct and safe use.

Opportunity for improvements:
- The designer should consider whether it would be feasible to accommodate other sticky tape roll sizes for use in the dispenser.
- Consideration should be given to methods of holding the dispenser securely in position, on a desktop, without increasing the overall weight of the product.

163. Interpreting numerical data

To calculate the average life expectancy of each chain, simply add up the life cycle hours and divide by the number of tests for each chain:

e.g. Chain A: 19 + 20 + 20 + 20 + 20 + 21 + 18 + 21 + 19
= 178 total hours
178 hours divided by 9 tests: 178/9 = 19.7 hours
Therefore chain A has an average life expectancy of 19.7 hours.
A = 19.7 B = 8.2 C = 14.1 D = 32.7 E = 10.7

164. Producing design ideas 1

Learner's own ideas.
Checklist: To make a contribution to the total of 9 marks available for a range of ideas, you must:
- Produce an idea that adequately addresses the requirements of the brief.
- Produce clear sketches with easy to follow annotations.
- Use technical terms in your annotations.
- Produce an idea that would work in the real world.
- Suggest types of suitable material, suitable manufacturing methods, suitable engineering processes and perceived advantages/disadvantages of the solution.

165. Producing design ideas 2

Learner's own ideas. A suitable sketch with basic information given; for example:

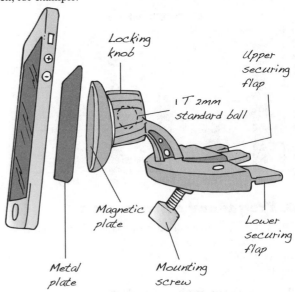

166. Modified product proposal

Learner's individual design proposal.
Showing that you can optimise your design through a series of realistic design proposals will contribute to the 30 marks available for the overall development section of your assessment.
Your designs for the stove legs should be easy to understand, clearly communicated, and be realistically achievable. You should also list the advantages and disadvantages of each proposal. You also need to justify your choice for the final design, and make informed decisions about material choice and manufacturing methods.

167. Justifying a design

Justifications might include:
I have chosen the bottle opener design because it fulfils the requirements of the design brief better than other designs I have explored.
- It is straightforward and cost effective to manufacture.
- It is safe to use as I will specify the removal of all sharp edges.
- It will open all bottles with a crown cap.
- It will aesthetically match other kitchen / bar items.
- It could be laser cut or blanked accurately in large quantities, for mass production of the product.
- Stainless steel does not rust and is therefore suited to use with foodstuffs.
- It can be stainless steel to enhance aesthetics and longevity.
- Stainless steel is resistant to bending, which is useful in the application of a bottle opener.

168. Justifying materials and processes

Answers might include:
Aluminium has low density (approx. 1/3 of steel), excellent strength to weight ratio, easy to machine, excellent corrosion resistance, easy to recycle, does not become brittle at low temperatures, extrudes easily into tubes, can be welded using traditional techniques, such as gas, TIG, MIG, friction welding etc., ideal for cycle frames.

169. Developing sustainability

Answers might include:
Power source is natural sunlight; product is small but effective, thus keeping material use to a minimum; packs very small, which keeps packaging to a minimum; plastic can be recycled; designed for longevity; provides light without using non-renewable energy sources.

170. Validating a design proposal

Validations for choosing the final design proposal might include:
- Provides multiple door-opening positions
- Simple but reliable
- Heat-resistant handle
- Positive closure
- Easy to manufacture

171. Evaluating with tools and techniques

Positive features could be:
Easy to manufacture, sturdy and strong, a cost-effective solution.
Negative features could be:
Not adjustable for uneven ground; do not fold away, thus making transportation difficult; the stove may be equally stable on three legs, which would be more cost effective.

Unit 6: Microcontroller Systems for Engineers

172. Microcontrollers and Unit 6

Individual responses for priority areas of revision.

173. Comparing different microcontrollers

Individual revision of important properties of the chosen microcontroller.

174. Project boards

Answers may include:
- Development of the system will be quicker, as you do not have to connect components.
- You do not have to consider interface requirements, as they are taken care of.
- Size is reduced because connections between components can be on the printed circuit board (PCB).
- The circuit may be more reliable, as there will be fewer temporary connections.

175. Using flowcharts

For example:

Symbol	Meaning	Symbol	Meaning
	Manual input		Manual operation
	Stored data		Start of a loop
	Internal storage		Parallel mode
	Display		Off page reference

176. BASIC as a programming language

Comments – convey information to a reader
Commands – instructions that the program carries out
Variables – used to hold values that the program manipulates
Labels – identifies parts of the program
Constants – fixed values

177. C as a programming language

```c
// This is an example of C program
//It will be used to highlight common features of the language

void setup() {
    // set the LED pins as outputs, the for() loop reduces the number of lines needed
  for (int pinNumber = 2; pinNumber < 5; pinNumber++) {
    pinMode(pinNumber, OUTPUT);
    digitalWrite(pinNumber, LOW);
  }
}
void loop() {
// This reads the ADC on pin A0 and assigns the value to the integer variable called sensorVal
    int sensorVal = analogRead(A2);
    int lowValue = 50;                      // This defines the integer value below which LED 2 will turn on
    int middleValue =200;                   // This defines the integer value above which LED 4 will turn on

// This tests to see if the ADC value is less than 50
if (sensorVal < lowValue) {
    digitalWrite(2, HIGH);            // This turns on LED 2
    digitalWrite(3, LOW);             // This turns off LED 3
    digitalWrite(4, LOW);             // This turns off LED 4
    }

// This tests to see if the ADC value is equal to or more than 50 and less than 200
  else if (sensorVal >= lowValue && sensorVal < middleValue) {
    digitalWrite(2, LOW);             // This turns off LED 2
    digitalWrite(3, HIGH);            // This turns on LED 3
    digitalWrite(4, LOW);             // This turns off LED 4
  }

// This tests to see if the ADC value is equal to more than 200
  else if (sensorVal >= middleValue) {
    digitalWrite(2, LOW);             // This turns off LED 2
    digitalWrite(3, LOW);             // This turns off LED 3
    digitalWrite(4, HIGH);            // This turns on LED 4
  }

  tone (6,200,100);
  }
```

178. Input and output devices

Example answer:
The sensor compares the temperature of the environment to that set on the thermostat. If the environment temperature is below the thermostat, the heating turns on. If the temperature is above that of the thermostat, the heating turns off.

179. Switches

For example:
Tilt switch – used as a safety feature to cut power to a device if it tips over (e.g. a portable electric fire) or used as an anti-tamper alarm in devices such as vending machines.
Micro-switch – used as a safety interlock on machine guards, or used as a burglar alarm trigger on doors.

180. Visible and infrared light-sensing devices

Answers must relate the change in light level to an output and explain its advantage to the user.
For example:
Mobile phones and tablets adjust their brightness levels, depending on the ambient light. This means that the screen is readable in bright light and that the battery use is reduced when dark. This makes it easier for the user to read the screen.

181. Temperature and humidity sensors

Specifications of thermistors, temperature and environment sensors to be located.

182. Input interfacing requirements

Answers should include examples of entertainment media: TV, CD, DVD, Blu-ray, mobile phones, internet, games machines, DAB.

183. Ultrasonic and control potentiometer

Answers may include, for example:
It is easier to tell the position of the slide potentiometer than the rotary one.
The slide potentiometer would take up more space than the rotary one.

184. Optoelectronic output devices

Answer must include a method used to generate a message on an LCD.

185. Electromechanical output devices

Answers may include, for example:
Relays – commonly used to control electrical devices in cars, such as windscreen wipers. Used to switch on and off outputs for high power audio amplifiers.
D.C. motors – children's toys, electric clocks
Servos – steering on remote control cars, DVD player to extend disc tray

186. Audio output devices

Answers may include, for example:
Buzzer – alarm clock, children's toys
Siren – burglar alarm
Piezo transducers – greeting cards, mobile phones
Speaker – TV, radio etc.

187. Transistor output stages

Gain = C/B = 500/0.7 = 714

188. Selecting hardware: input devices

Individual responses. Investigations must include specifications for one of the components listed on page 189.

189. Selecting hardware: output devices

Individual responses. Investigations must include specifications for one of the components listed on page 190.

190. Generating a system design

Any five examples from the following:

	Reed switch		Crystal
SW2 1W		X1 32 768kHz	
BL1	Signal lamp	TR MD BC IN1 150 cm	Ultrasonic range finder
Q1 500 lux	Photo transistor	Q2	Opto-isolator
MT1 M	Motor	MT2 M	Servo motor
RL1	Relay	DS1	7-segment display
IN4 DS1990A	iButton	D1	Flashing LED

191. Safe use of typical electronic tools

For example:
The wire cutters could cause the cut-off part of the component to fly out. Wear goggles.
The wire strippers could cut fingers. Ensure fingers are kept away from the cutting edges.
The pliers could cause crush injuries. Keep fingers away from the jaws.
The screwdrivers could cause puncture injuries. Rest the object being worked with on a bench. Do not hold in the hand.

192. Assembling and operating a microcontroller system

Any three example inputs from:
Heat sensor
Light sensor
Touch sensor
Infra-red / ultrasonic receiver

Any three example outputs from:
Buzzer
Speaker
Electric motor
Relay
LCD
7-segment display

193. PICAXE® and Logicator: program files and error checking

Answers should include a description of syntax error and detection. For example:
A syntax error means that the way a program has been written does not conform to the required format. For example, the use of a lowercase letter when an uppercase letter is required. Syntax errors can be detected in several ways. For the Microchip IDE errors are detected in an interactive manner. This means as soon as a syntax error is entered by the system designer, the software highlights where the error is and suggests possible solutions. Another method is to wait until the program is complete and the system designer either chooses to check the program for errors or tries to send the program to the microcontroller.

194. PICAXE® and Logicator: simulation, compiling and debugging

Example answer:
Software compiling is the process of translating high-level programming languages that humans can read into the low-level instructions that microcontrollers can perform.

195. Microchip, PICkit 3 and MPLAB®: program files and error checking

Example answer:
When a program has multiple errors each one can impact on the other. This interaction can mean it is very difficult for the system designer to follow the cause and, hence, solution of the problem. By resolving any problems as soon as they arise the problems are not compounded, making resolution easier.

196. Microchip, PICkit 3 and MPLAB®: simulation and debugging

Example answer:
The PICkit 3 hardware removes the need for a computer. This means that microcontrollers can be programmed in environments where a computer might get damaged, or not be able to work. As exactly the same program will be sent, the user can ensure no changes have taken place. This could make the program more secure. The process of transferring the program will probably be quicker than when doing so through a computer, therefore the system designer will be more productive.

197. GENIE®: program files

Example answer:
Advantage:
The software translates flowchart symbols into code that can be sent to the microcontroller. This means the system designer does not have to worry about the syntax of the language, hence reducing this type of problem.
Disadvantage:
Flowchart symbols can take up more space than a text-based program. For more complex programs this can mean that the system designer may find it difficult to follow the entire sequence of operation.

198. GENIE®: simulation

Example answer:
Simulating the program before downloading it to the microcontroller saves time. If the program does not perform as expected modifications can be made. The simulation will be free from hardware defects. For example, the hardware may have an LED that does not work. If the program is downloaded to the microcontroller and the LED does not operate, the fault could be with either the hardware or software. By simulating the program before use, a potential software cause of the problem can be eliminated.

199. GENIE®: compiling and debugging

Example answer:
The Debug Live system allows the user to view real-time information about the devices connected to the microcontroller. This allows input / output actions to be accurately calibrated to perform as required. Potential unexpected interactions between components can be observed and appropriate adjustments made if needed.

200. Arduino™ Uno: program files and error checking

Example answer:
A library is an additional set of functions / commands that extend the capability of the ones built into the programming environment. For example, a manufacturer of a modular sensor can supply users with routines to make it easier to use the sensor.

201. Arduino™ Uno: simulation and testing

Example answer:
Interactive simulations are not built into the Arduino™ platform. This means that a separate piece of software will be required. There are a range of free options. In order to use these, the system designer assembles the required hardware virtually on-screen, enters or loads the program and observes the outcomes.

202. Flowcode and E-block: creating and managing program files

Example answer:
Each type of microcontroller will have specific performance capabilities; for example, the number of inputs or its speed of operation. Depending on the situation the microcontroller is being used in, the system designer may need to select one from a particular platform. As Flowcode works over multiple platforms, the system designer will have to develop only one set of skills and knowledge to be able to make use of more than one platform. This increases their productivity.

203. Flowcode and E-block: simulation

Example answer:
Advantage:
A 2D or 3D representation clearly shows people what the system will do. This means a person not familiar with microcontroller development will be able to understand the operation of the system without having any prior programming knowledge.
Disadvantage:
The system depends on having models of the input or output device to display. This may mean some devices will not be available to the user.

204. Flowcode and E-block: simulation and debugging

Example answer:
An oscilloscope provides a graphic display that shows how a voltage changes over time. The user can determine the maximum and minimum voltages to be displayed, as well as how frequently the display updates. When required, the display can be paused to allow fast-changing events to be viewed.

205. Coding practice and efficient code authoring

Individual responses. Answers must include:
File name
A description of the purpose of the file
A list of the files used by the program
Who the program was written by
When the program was written
Where the program was written
The version number of the program
How the program has been updated

206. Coding constructs: inputs and outputs (BASIC and C) 1

Answers must include one situation where the format of the PICAXE® command would be easier to use and one where the format of the Arduino™ C command would be easier to use.
Example answer:
In the PICAXE® command, the user enters the information for all eight bits, even if only one is going to be turned on. This means the command will be good for working with multiple bits, but is probably more complicated than the Arduino™ command when working with single bits. For example, in PICAXE® if the user wants to turn on only pin 0, the following command would need to be entered:

Let pinsA = %00000001. The user will need to know which digit relates to which output pin. From left to right, the first digit links to pin 7, and the last to pin 0. For the Arduino™ the pins are explicitly referred to, e.g. *pinMode(13, OUTPUT)*, which makes understanding their function easier. However, if several pins needed to be turned on the command would have to be entered multiple times; for example:

pinMode(13, OUTPUT);
pinMode(12, OUTPUT);
pinMode(11, OUTPUT);

The more lines that need to be entered into the program the greater the risk for errors to occur.

207. Coding constructs: inputs and outputs (BASIC and C) 2

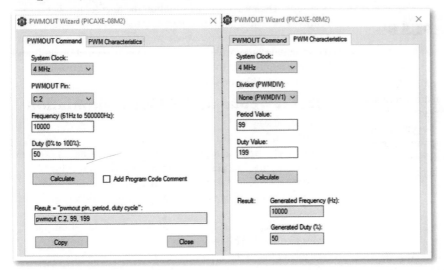

The PWM Wizard generates the appropriate code to copy and paste into the program by entering the desired values into menu boxes. This means the code produced, shown above in the Result box, will be free from syntax errors. This method will reduce the time it takes to develop a functional program.

208. Coding constructs: inputs and outputs (GENIE® flowchart) 1

Example answer:
Advantage:
This makes it easier for the system designer to use, as they do not have to configure the pins to be inputs or outputs.
Disadvantage:
This makes the microcontroller less flexible compared to other types of microcontroller.

209. Coding constructs: inputs and outputs (GENIE® flowchart) 2

Using GENIE® to construct a flowchart that sends a single pulse of 100 ms duration:

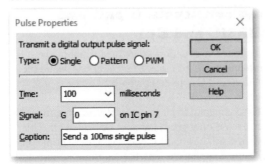

210. Coding constructs: logic and arithmetic variables and arrays

PICAXE® makes use of the table command. This allows the system designer to store byte constants (0–255) in an array-like structure when the program is downloaded onto the microcontroller. These values cannot be changed as part of the program.
The Genie® microcontroller uses the commands **put** and **get** to enter and retrieve data from an array. The format of the put command is put value, index
The format of the get command is get result, index

211. Coding constructs: logic and arithmetic

Individual answers, which must include AND, OR and NOT truth tables for a range of input conditions.

212. Coding constructs: program flow and control 1

Example answer:
The void key word is used only in function declarations. It indicates that the function is expected to return no information to the function from which it was called.

213. Coding constructs: program flow and control 2

Individual answers, which must include how delays can be achieved on a chosen platform.

214. Program flow: iteration

Individual answers, which must explain the differences between *loop while* and *loop until* on a chosen platform.

215. Control of program sequence: *if else*

Individual answers, which must explain the use of the *if else* flow control functions on a chosen platform.

216. Control of program sequence: *switch 1*

```
if condition then
    {code}
    elseif condition then
        {code}
    else
        {code}
    end if
```

Instead of an 'else' command it is possible to use an 'elseif' command and specify another condition. If the first condition is true, the code between the 'if condition then' and 'elseif condition then' command will be executed, otherwise the 'elseif' condition will be evaluated and, if true, the code between 'elseif condition then' and 'end if' will be executed.
An 'else' command can be added to an 'if' and 'elseif' command sequence, which will be executed if no previous 'if condition then' or 'elseif conditon then' has evaluated as true.

217. Control of program sequence: *switch 2*

Like **if** statements, **switch...case** controls the flow of programs by allowing programmers to specify different code that should be executed in various conditions. In particular, a *switch* statement compares the value of a variable to the values specified in *case* statements. When a *case* statement is found whose value matches that of the variable, the code in that *case* statement is run.
The **break** key word exits the *switch* statement, and is typically used at the end of each *case*. Without a break statement, the *switch* statement will continue executing the following expressions ('falling-through') until a break, or the end of the *switch* statement is reached.

218. Structured program design

Individual answers, which must include a flowchart to run an alarm system.

219. Number systems: decimal to binary

Decimal to binary: 10000011, 10101100, 01011110

220. Number systems: binary to decimal

Binary to decimal: 141, 62

221. Project analysis

Individual's own design brief that includes requirements and restrictions.

222. System design and program planning

Individual answers, which should identify input/output components to be familiar with. An example response may be 'I have not used an electric motor before, this is something I will investigate.'

223. System assembly, coding and testing

Answers should explain why coloured wires are an advantage. For example: On a typical breadboard circuit, wires will be placed very close to each other and may overlap. By having the wires a different colour, the system designer will be able to follow more easily where they start and where they end. The only problem with this is there is a limit to the range of colours available.

224. System testing and operation

Individual answers, which should include a feature of a system developed and a description of its operation.

225. Production of evidence

Answers should be in the form of a simplified guide, such as this example:
1 Produce a technical specification
2 Produce a test plan
3 Select and justify input and output devices
4 Define system connections
5 Produce initial design for program structure
6 Produce annotated copies of code
7 Record test data and analyse
8 Record audio-visual of system operation
9 Maintain a structured project log throughout

226. Your Unit 6 set task

Your notes on the Unit 6 set task, always referring to the latest Sample Assessment Material on the Pearson website for an indication of assessment details.

227. Reading a brief

Example answer:
Safety would be a key factor when a product is used by children. Depending on their age risks, such a swallowing, small parts could be significant.
The ergonomics of the product should take into account the anthropometry of the intended users.
Interactions between the users and the control systems of the product should be of an appropriate complexity for the target age of the users.

228. Creating a task plan

The time needed for each of the stages will be in response to information on timing and marks from the Pearson website.

229. Monitoring progress

Example sensor list:
Active buzzer
Passive buzzer
Vibration switch
Photo resistor
Tilt switch
Infrared emission sensor
Temperature sensor
Infrared sensor
Microphone
XY-axis joystick module

Example output list:
Piezo sounder
Loudspeaker
Servo
Electric motor
Stepper motor
LCD screen
Range of size and colour LEDs

230. Monitoring changes 1

Example answer:
From version 1 of the program, lines 42 and 43 have been broken into two parts, as shown by lines 45–48 in version 4. The change to the structure makes the program easier to follow for the reader. In version 2 of the program, an additional output, C.5, was included. It was discovered that this command was duplicated elsewhere and therefore was not required. It is removed from versions 3 and 4.

Version 3 of the program includes two 5 second delays. Again, these were duplicated in other parts of the program and were not needed.

The final version of the program performs as required and would be clear for a competent third party to read.

231. Monitoring changes 2

Example answer:

The text on the switch panel (yes/no and question answered) is obscured by the user's hand. If the text was moved above the switches and the user operated the switches from below it would make it clearer to operate.

Where text currently reads Answer, Correct, Incorrect it would be clearer if the text read correct answer and incorrect answer.

Instead of the text saying players it could be amended to 'The light indicates which player has answered the question'.

All of these modifications would enhance user experience for an improved response.

232. Analysing a brief for product requirements

Example answer:

A buzzer or piezo sounder would give an audible indication of when the time is up.

An LED would be a simple method to provide a visual indication of the time being up.

233. Completing a test plan

Example answer:

Currently, the system uses a sound to indicate when the 10 s have elapsed. The players will get no warning regarding how much time has elapsed until the time is up and then it would be too late for them to answer. A system could be incorporated that would count down the time remaining. For example, a LCD display could display a countdown from 10 to 0 seconds. This would prompt the player to respond before the time has run out.

234. Formulating a system design

Example answer:

I will use LEDs to indicate player responses and outcomes. LEDs are low cost, which will allow me to use my allocated budget for other things. As LEDs require a low current they can interface directly with the microcontroller. My centre also has LEDs in stock, which means they are available for me to use without any delay.

235. Proposing system connections

Example answer:

I would add a label to each LED that names the player the LED is associated with. For example, D1 could be labelled 'Player 1 LED'.

236. Planning a program structure

Example answer:

The program structure does not consider what would happen if no player pushed their button. Another situation the program structure does not consider is what would happen if a player pushes their button but then does not provide an answer that is either correct or incorrect.

237. System assembly and programming

Example answer:

```
1
   if pinC.0=1 then              'Has player 1 pushed their button
   High B.3                      'If they have light their LED
   gosub ten_seconds_answer      'Give them 10 seconds to answer
   end if                        'Move to next player

   if pinC.1=1 then              'Has player 2 pushed their button
   High B.2                      'If they have light their LED
   gosub ten_seconds_answer      'Give them 10 seconds to answer
   end if                        'Move to next player

2

   ten_seconds_answer:           'Allows player 10 seconds to answer
   pause 100                     'Waits 10 seconds
   if pinC.4=1 then              'wait for the quizmaster to turn on the buzzer
   high B.5                      'Sounds a buzzer when the time to answer is up
```

238. System testing

Example answer:

1   An audible signal could be produced by a loudspeaker, buzzer or piezo sounder. If each player is associated with a different frequency, a single source of sound could be used for all players.

2   This code turns on the correct indicator light, as required by the design brief, to show which player has pushed their button. It is clear and easy to read the output indicator.

```
if pinC.0=1 then 'Has player 1 pushed their button

High B.3_____'If they have light their LED
```

239. Results analysis

Example answer:

A possible unexpected event that could occur would be failure of one of the LEDs, possibly due to excessive current. In order to test this, a working LED could be replaced with one that was broken. Then, by pushing the appropriate player button linked to the LED, the system's response could be observed.

240. Recording a system in operation

Example answer:

Recording of the evidence could be paused during the 'waiting' time. If the video recording included a clearly visible timer, there would be sufficient evidence to show the critical events; i.e. the buttons being pushed and the correct / incorrect LED coming on without the wait being recorded while nothing was happening. Another approach would be to record the video in real time and edit out the waiting time between each event.

241. Recording a commentary

Example answer:

When player 1 activates their button the voltage on input pin C.0 will change from 0 volts to +5 V. The microcontroller has been programmed to respond to this change. Until this change occurs the program will loop around, testing the state of the four players' input pins. The *if* command performs this test.

When the button is pressed the output on pin B.3 will change from 0 V to 5 V. This will energise the LED connected to this pin. This informs the players and quiz master that player 1 was the first to respond.

As the next section of the program that runs is common to all players, it has been written as a subroutine called ten_seconds_answer. This is an efficient way in which to structure the program. This subroutine waits for 10 seconds before setting the output of pin B.5 high. As the buzzer is connected to this pin it will begin to sound. The program will keep the output high for 2 seconds and then turn it off. The wait command has been used to set the duration to 2 seconds.

From this subroutine the program goes to the Quiz_Master subroutine. This subroutine loops around until the quiz master indicates if the player's answer is correct or incorrect. If the voltage on pin B.6 is high the program interprets this as a correct answer and sets the output pin B.6 high. This is the correct answer LED indicator. If the voltage on pin B.6 is low the program interprets this as an incorrect answer and sets the output pin B.7 high. This is the incorrect answer LED indicator. These actions meet the demands of the client brief.

At this stage the design and implementation of the question answered button does not function in an ergonomically efficient manner. This is not a significant issue as, given time, a better solution could easily be designed and installed.

Published by Pearson Education Limited, 80 Strand, London, WC2R 0RL.

www.pearsonschoolsandfecolleges.co.uk

Copies of official specifications for all Pearson qualifications may be found on the website: qualifications.pearson.com

Text and illustrations © Pearson Education Limited 2017
Typeset and illustrated by Kamae Design, Oxford
Produced by Out of House Publishing
Cover illustration by Miriam Sturdee

The rights of Andrew Buckenham, Kevin Medcalf, David Midgley and Neil Wooliscroft to be identified as authors of this work has been asserted by them in accordance with the Copyright, Designs and Patents Act 1988.

First published 2017

20 19 18 17
10 9 8 7 6 5 4 3 2 1

British Library Cataloguing in Publication Data
A catalogue record for this book is available from the British Library

ISBN 978 1 292 15028 4

Acknowledgements
The author and publisher would like to thank the following individuals and organisations for permission to reproduce photographs:

(Key: b-bottom; c-centre; l-left; r-right; t-top)

123RF.com: 151, Olga Serdyuk. 135t; **airbus.com:** 101t; **Alamy Stock Photo:** Action Plus Sports Images 96, Adrian Sherratt 128t, age fotostock 104, Aviation Images 131c, Carmel Auad 133br, Caryn Becker 131t, Chris Wilson 128c, 133tl, Clynt Garnham Industry 131b, Dejan Jekic 117t, dpa_Picture_Alliance 119l, GIPhotoStock X 57l, imageBROKER 118l, INTERFOTO 132l, John Crowe 135b, NaturaLight - No Release Needed 156br, Oleksiy Maksymenko Photography 99, 132r, Piero Cruciatti 120l, Richard Heyes 168b, scottishcreative 25, Seapix 102t, speedix 103, Stuart Kelly 128b, Tony Watson 109, Universal Images Group North America LLC 107, Zuma Press Inc 169; **American National Standards Institute (ANSI):** 129bc; **BSI:** 129tl; **European Standards Organisation:** 129cl; **Fotolia.com:** canonlife 162, frog 116br, geargodz 123, Hamik 127, Hellen Sergeyev 28, icarmen13 106, sabuhinovruzov 133bl, shara 97; **Fraunhofer-Institut für Keramische Technologien und Systeme IKTS:** 120r; **German Institute for Standardisation (DIN):** 129tr; **GESIPA ®:** 118b; **Getty Images:** Antoine Gyori 98t, David Boyer 121t, John Burke 168t, lleerogers 213, mhaperrosa 98b, RickLeePhoto 108, Science & Society Picture Library 114, sspopov 136; **Honeywell Ltd:** 115; **ISO:** 129br; **Kevin Medcalf:** 174t, 174cl, 185br, 190t, 190c, 190b, 192c, 231, 238t, 238b, 239, 241t, 241b; **Laurent Ballesta:** 113b; **Lisin Metallurgical Services LLC:** 111; **Lone Star Hovercraft:** 112; **Matrix Technology Solutions:** 174c, 174bl; **NASA:** 8, 121b; **Pearson Education Ltd:** Gareth Boden 133tr, Coleman Yuen. Pearson Education Asia Ltd 101b, Jules Selmes 181t, Rob Judges 139, Tsz-shan Kwok 167; **Rapid Education:** 71, 174br, 178, 179 (a), 179 (b), 179 (c), 179 (d), 179 (e), 179 (f), 179c, 179cr, 179b, 180cl, 180b, 181c, 181b, 183t, 183bl, 183br, 184t, 184tr, 184c, 184br, 185tl, 185c, 185bl, 186tl, 186tr, 186c, 186b, 187, 191 (1), 191 (2), 191 (3), 191 (4), 191 (5), 191 (a), 191 (b), 191 (c), 191b, 192t; **Science Photo Library Ltd:** Alistair Philip 105, Ben Johnson 116t, James King 117b; **SETFORGE ENGINEERING:** 122; **Shutterstock.com:** 630475 180t, Ambient Ideas 119b, Andrey Eremin 116b, AummGrapixPhoto 116c, Chicco DodiFC 100, Christian Lagerek 126, Evgeny Korshenkov 119r, eye idea 134, Gareth Boden 60, kurhan 102b, Mrs_ya 274, Oleksiy Mark 125, Smileus 57r, Stu49 180cr, Sue_C 16, Yegor Korzh 39; **Volt Planet:** 185tr; **Wikimedia Commons:** Stahlkocher 113t; **www.picaxe.com:** 174cr, 192b; **Neil Wooliscroft:** 156t, 156c, 165

All other images © Pearson Education

We are grateful to the following for permission to reproduce copyright material:

Page 10, Harry Smith, Mathematics C1 C2 M1 S1 D1 Revision Guide, Pearson Education Limited. © 2013. Pearson Education, Inc; page 48, © Adrio Communications Ltd, Used with permission; page 55, HyperPhysics, © 2012 C.R. Nave. Used with permission of Author. http://hyperphysics.phy-astr.gsu.edu/hbase/thermo/carnot.html; page 141, Used with permission of The Greenage. http://www.thegreenage.co.uk/wp-content/uploads/2013/12/UK-Energy-Mix-2013.png; pages 175, 184, 187,197, 198, 199, 208, 209, 212, 256, © New Wave Concepts Limited. All rights reserved; pages 177, 181, 192, 200, 201, Used with permission of The Arduino AG; pagea 193, 194, 205, 235, © PICAXE www.picaxe. com, Used with permission of The PICAXE; pages 195, 196, ©Microchip Technology Incorporated. All rights reserved; pages 202, 203, 204, 205, © Matrix Technology Solutions Ltd 2017. Used with permission; page 253, © CS Odessa Corp

Notes from the publisher

1. While the publishers have made every attempt to ensure that advice on the qualification and its assessment is accurate, the official specification and associated assessment guidance materials are the only authoritative source of information and should always be referred to for definitive guidance.

Pearson examiners have not contributed to any sections in this resource relevant to examination papers for which they have responsibility.

2. Pearson has robust editorial processes, including answer and fact checks, to ensure the accuracy of the content in this publication, and every effort is made to ensure this publication is free of errors. We are, however, only human, and occasionally errors do occur. Pearson is not liable for any misunderstandings that arise as a result of errors in this publication, but it is our priority to ensure that the content is accurate. If you spot an error, please do contact us at resourcescorrections@pearson. com so we can make sure it is corrected.

Websites

Pearson Education Limited is not responsible for the content of any external internet sites. It is essential for tutors to preview each website before using it in class so as to ensure that the URL is still accurate, relevant and appropriate. We suggest that tutors bookmark useful websites and consider enabling students to access them through the school/college intranet.